Wild Track

Wild Track

Sound, Text and the Idea of Birdsong

Seán Street

BLOOMSBURY ACADEMIC
NEW YORK • LONDON • OXFORD • NEW DELHI • SYDNEY

BLOOMSBURY ACADEMIC
Bloomsbury Publishing Inc
1385 Broadway, New York, NY 10018, USA
50 Bedford Square, London, WC1B 3DP, UK
29 Earlsfort Terrace, Dublin 2, Ireland

BLOOMSBURY, BLOOMSBURY ACADEMIC and the Diana logo are trademarks of
Bloomsbury Publishing Plc

First published in the United States of America 2023
This paperback edition published 2025

Copyright © Seán Street, 2023

Cover design: Louise Dugdale
Cover image: Private Collection Accademia Italiana, London / Bridgeman Images

All rights reserved. No part of this publication may be reproduced or transmitted in any form or by any means, electronic or mechanical, including photocopying, recording, or any information storage or retrieval system, without prior permission in writing from the publishers.

Bloomsbury Publishing Inc does not have any control over, or responsibility for, any third-party websites referred to or in this book. All internet addresses given in this book were correct at the time of going to press. The author and publisher regret any inconvenience caused if addresses have changed or sites have ceased to exist, but can accept no responsibility for any such changes.

Whilst every effort has been made to locate copyright holders the publishers would be grateful to hear from any person(s) not here acknowledged.

A catalog record for this book is available from the Library of Congress.

ISBN: HB: 978-1-5013-9794-3
PB: 978-1-5013-9798-1
ePDF: 978-1-5013-9796-7
eBook: 978-1-5013-9795-0

Typeset by Newgen KnowledgeWorks Pvt. Ltd., Chennai, India

To find out more about our authors and books visit www.bloomsbury.com and sign up for our newsletters.

For Jo

Touch lightly Nature's sweet Guitar
Unless thou know'st the Tune
Or every Bird will point at thee
Because a Bard too soon
                    – Emily Dickinson

Contents

Prelude												viii

1 Dawn chorus: Describing sound							1
2 Recording an essence: Hearing, reading and the imagination		19
3 Ludwig Koch and the music of nature						35
4 Murmurings: Call and response between bird and human			53
5 Paths through the green wood							71
6 Science in Arcadia: The road to Selborne					91
7 The romantic poetry of listening						111
8 Honest John: The sound world of John Clare					129
9 North American sublime								147
10 Survival and the sound of spirit							167
11 Listening to ourselves listening: Voices for today and tomorrow		183

Postlude: 'Adieu! adieu! thy plaintive anthem fades'					201
Bibliography											207
Index												215

Prelude

The surface of the Earth is finite; everything that has ever been recorded – through technology or before that, through words or images – has been experienced somewhere, in a place where a human being once stood or sat. That place still exists, however changed, and many of the first sounds also remain the same: rain falling on stone, waves breaking on shores and the wind through woodland trees. Someone, perhaps millennia ago, was where I am now, and heard the world going on around them as I do now, measuring the moment in the place, listening. It is an act of the imagination to picture how a square mile of land has evolved, adapted, been manipulated and shaped, yet beneath it all, has remained fundamentally itself. A place is geographically circumscribed, but time is thorough and ongoing. Furthermore, to extend the concept of continuous sound evolution to our everyday footpaths, suburban roads and parks, the street corners and retail centres of today and hear them as they were, may be hard, but potentially salutary. A shopping mall's tarmac car park may have once been marshland where wildlife thrived, a housing estate may have wiped out meadowland, and so on.

Some things sounded differently here in the past, some did not. Like travelling through a once familiar city, finding it changed in some respects, but with flashes of familiarity as certain scenes evoke memories, there may still be constants. Imagination and concentration can seed the idea of past sound, and words have the capacity to express it, and sometimes that same imagination moves beyond documentation into a kind of transcendence.

This book is entirely about listening as an active occupation: conscious listening, awareness, attention to sounds under sounds, all the noises and murmurs of the shouting and whispering world. While it might have selected another sound source to make its case, it happens to have as its focus birdsong within the context of the natural world, because in various forms, it is all around us, and has long been a constant through the generations, so we have in its presence a reference. The songs of birds have preoccupied human beings since they first heard something they did not understand and sought to interpret it. The very folk names that have passed down to us through time are often, themselves, attempts at simulating their music, and the fundamental and primary impetus of recording the aural experience of birds in words came from the sound itself; an onomatopoeic interpretation of the call as it impressed itself on human consciousness, so in many cases, identification led to naming according to an approximation of the sound heard. Some bird names are buried in Time, while some are more self-evident.

Time, indeed, is part of the equation: like us, birds live and die, leaves fall, but the seasons, the blossoms and the songs return year after year, or at least, they have up to now ... We listen and we philosophize, but the birds don't: they sing, unconcerned with our witness below. They do what they have always done, and we cannot help but try to find our own meanings in their song. The Jesuit Gerard Manley Hopkins wrote of the very eggs of a thrush as 'little low heavens',[1] but those heavens are part of the real world, as we are. Birds live and die, they fight and kill, they defend territory, scavenge and try to protect and feed their young, just as we do; it's just that they don't discuss it with us. Neither can we emulate their avian abilities; to take to the air, we must build machines. We can at least sing, we can make our own kind of music, but we learnt that from them in the first place too.

Birdsong seems to take us somewhere far, far back, to somewhere we can almost remember. It is a hint of a lost sound world re-emerging, a metaphor for the planet as it was before we began obliterating it, before the air changed, when the light was clearer. At the moment, it feels as though nothing is different; the new leaves and blossoms come on our trees, and from within them, often unseen, the birds sing. Location-based sound recordists and producers frequently gather ambient sound from the air which they call 'Wild Track', the ambience of the world happening around us, collected as part of the production process, usually to blend with other sounds and the spoken word, to smooth editing and create a continuity of atmosphere. Latterly, however, partly because of pandemics and resulting lockdowns, the background has become more of the foreground in our philosophy of sound, and the wash of tonal colour onto which the main subject was placed now assumes its own importance, and so we listen to what it has to say with retuned ears, minds and hearts. The wild track is its own document.

This started me thinking about the very act of recording: how we take for granted the fact that technology can provide the sounds we want to hear at the

---

[1] G. M. Hopkins, 'Spring' in *The Poetical Works of Gerard Manley Hopkins*, edited by N. Mackenzie (Oxford: Clarendon Press, 1990), p. 42.

touch of a button. On my phone there is an app that records the song of a bird as an aid to identification, websites offer me illustrated articles that inform, using recordings, and I can evoke whole landscapes through surround sound, simply by setting up a few speakers linked to playback. All this I accept. But it prompts a question: what is the consolation for the nature lover, faced suddenly, or gradually and inexorably, with the absence of all this, say through the loss of the faculty of hearing? From this leads another question: how did we share the experience of the sound environment around us before the technology we now employ so easily and readily? For someone without access to the sound of the world, how did – and do – you name, describe and *explain* these sounds? It is when we try to describe what we hear from them, that we must resort to words. That quest is of course not only a response to birdsong, but to the whole natural world of which we are a part. This book aims to join the listener in that wider world, but in doing so, it confronts the sheer scale of such a task. So, we come back to minutiae, the details from which our sound environments are made and build themselves. Birdsong lends itself to this focus, because as with human expression, it is communication, it is its own speech and language. Very little epitomizes the poignancy of the disappearing moment more eloquently than the song of a bird.

We continue with what we have always done: we seek to translate it and find metaphors of sound that we can interpret in our own terms. Even the 2014 *Handbook of British Birds* published by the RSPB (Royal Society for the Protection of Birds) finds recourse in similes to convey the sound of, for example, the long-eared owl: 'jingling calls, but later they make a drawn-out squeak that has been likened to a squeaky gate,'[2] or the voice of the Jack Snipe, one of the wader species, occasionally making an in-flight call 'sounding like a cantering horse – "kollarap, kollarap, kollarap"'.[3] It is not hard, considering just these few examples, to gain an understanding of how the sound of birds, their identification and recognition, has, through history, been so often linked to an attempt to describe or explain to another, the sound of the bird itself. Because we are not birds, we mostly (with notable exceptions) lack the ability to convincingly mimic them. We fall back on our own language to interpret theirs, and from this comes a challenge that has fed into some of the greatest human forms of cultural expression. To 'hear' the soundscapes of our ancestors, how sound was recorded

---

[2] P. Holden and T. Cleeves, *The RSPB Handbook of British Birds* (London: Bloomsbury, 2014), p. 188.
[3] Ibid., p. 134.

as an experience, prior to the advent of recording technology, we must rely on words. This is an attempt to explore that premise, using the idea of birdsong and the natural world as its case study. It is not a naturalist's book; neither is it a literary guide, although it strongly and gratefully acknowledges both worlds. It seeks not to be comprehensive, but indicative. As far as the literary examples I have chosen are concerned, it inevitably comes down to a personal choice. It is also not a chronological survey of its subject; mostly it is an exploration of themes as they feel relevant to the subject under discussion, rather than how one thing led to another, although sometimes it's that too. After all, whether we are discussing the sonic world of nature with Pliny the Elder, who died seventy-nine years before Christ was born, or Dara McAnulty, the young Irish naturalist born in 2004, we find ourselves considering writers who marvel at the same phenomena, wondering how best to convey the experience of listening to it.

My subtitle is 'sound, text and the idea of birdsong'. In its own way of course, *everything* is text. Sound is a text written on air; a thought is a text written in the mind. For convenience, however, and with this caveat firmly in view, this book engages with the word 'text' mostly in its written sense, as the transference of ideas and sounds onto the written or printed page. So why should this matter? Why should I invite you to read poetry and creative prose in a book about sound and listening? It is my hope that some of the examples I cite here – and there are many others that could have been selected – may prompt the reader to delve further into these subjects, read, consider and, beyond all that, listen with a heightened awareness to the sounds of the world. If we hear a thing well enough to accurately describe it, to articulate its qualities and its effect upon us, then we are probably listening more deeply than we were before we focused our inner microphone on that sound. To *describe* a sound opens the door to truly *knowing* it. So, here's a thought for sound recordists: don't just let your machine do the listening but see how well you can do what the machine cannot: that is to say, capture not only the sound of a place or a thing, but what it felt like to be there at the time. I would suggest that an attempt – even a failed attempt – at putting down on paper our experiences of a sound in words – or voicing them into that same machine – will bring us closer to the idea of not only hearing but *knowing* sound. It may help us to hear and know one another better too.

It seems to me that the choice of tuning in to birdsong as a metaphor for all listening, presents itself as an ideal subject for this exploration; birds are there around us, and their sounds accompany us through our days, whether we notice them or not. Birds and other animals may have shown us the way towards speech, as they certainly continue to inspire us to sing; and most importantly, the planet is withering, and as it grows noisier in its self-destruction, the life that it first nurtured becomes quieter with each day that passes. It matters that we notice things now.

My sincere thanks to the first inspirers of this journey, and for their encouragement as I embarked upon it. I owe a lot to Cheryl Tipp, Curator of Wildlife and Environmental Sounds at the British Library, who was the first to encourage me to think that perhaps there was a case to answer here. That said, without the supportive words and positive advice of Hilary Davies, Tim Dee, Kevin J. Gardner, Jeremy Hooker, Richard Mabey and Chris Watson, I might not have had the temerity to set out on such a hazardous interdisciplinary voyage. In the end however, because this is a book about listening, if it helps in the understanding of how tuning more actively to the auditory sense leads to increased care and thought in the act of recording in sound and words, in making, creating and enhancing life, then it will have fulfilled its ambition. My thanks to kindred spirits Mike Collier and Michael Guida, and other colleagues and friends at Liverpool University, Goldsmiths, the University of London, the University of Sunderland and the John Clare and Richard Jefferies Societies. Jeremy Hooker's poem, 'Somewhere, a blackbird', from his 2019 collection, *Word and Stone*, is reprinted by permission of the author and the publisher, Shearsman Books. To my wife Jo, my love, and thanks for her reading of the text, and her valuable suggestions throughout the whole process.

1

Dawn chorus: Describing sound

Birdsong is a doorway into a parallel place. It is at once familiar and profoundly other-worldly, part of a way of being that we have sought to understand, and coveted, since we first listened, and looked up. I once spent a chilly early April morning in 1982, in the woods above Thomas Hardy's cottage in Higher Bockhampton, Dorset, listening to and trying to record the dawn chorus, hoping to capture something of the sound that he might have heard as a boy. In the end, I only partially succeeded in my efforts of course, because although it *was* a dawn chorus, it was not *Hardy's* dawn chorus: I was about 130 years too late for that. In spite of this, it was not just any wood; notwithstanding the time lapse, it remained Hardy's wood, and as it happened, for me it was not just any time, but a day near the start of April 1982, in the first week of the Falklands War. When I think of that morning, I remember the sounds I heard, the broader context of world events in which I listened, and in my imagination, Thomas Hardy's experience of it in his time, as evoked in the natural music that plays and weaves through the text of his novel, *Under the Greenwood Tree*:

> To dwellers in a wood almost every species of tree has its voice as well as its feature. At the passing of the breeze, the fir-trees sob and moan no less distinctly than they rock; the holly whistles as it battles with itself; the ash hisses amid its quiverings; the beech rustles while its flat boughs rise and fall. And winter, which modifies the note of such trees as shed their leaves, does not destroy its individuality.[1]

That is Hardy's 'voice', articulating his interpretation of what he heard on a particular day, filtered through reflection. We are listening to him listening, just as when we hear a recording, we are listening through a microphone, wielded by another. Often for the sound recordist, woodland – or indeed any open-air ambience – can be a mixed blessing. Wind noise through trees may be ambiguous; is that the sound of leaves and branches responding to the breeze,

---

[1] T. Hardy, *Under the Greenwood Tree* (London: Penguin Classics, 2004), p. 7.

or is it the sound of rushing water, or rain? The eye comes to the aid of the ear in the actual location, explaining the origin of the sound, but to the 'blind' listener, the mind asks questions the ear cannot answer. Likewise, in modern times, there are the everyday 'sound-bombs' of a distant car, an aircraft or a siren. No such problem for the written word, filtering and focusing like a spotlight on the centre of attention; Hardy's precision conveys the sound – give or take a century or so – of the wood in which I found myself that morning just as the birds woke up. His ears had already done the work, and it was his mind that interpreted what they heard, recorded it and then played it back through words to be inwardly 'heard' by the reader: in this case, me.

In May 1924, less than two years after the birth of the British Broadcasting Company (BBC), listeners heard the well-known cellist Beatrice Harrison playing popular solo pieces in the garden of her house, Foyle Riding in Surrey, among them *Songs My Mother Taught Me*, by Dvorak, *Chant Hindou* (Rimsky-Korsakov) and *The Londonderry Air*. Harrison was a familiar figure on concert platforms, the soloist preferred by Edward Elgar in performances of his *Cello Concerto*, and the first to record that particular work under his baton. Already a celebrity by the time of her garden recitals, it was not however Harrison to whom listeners responded in their thousands, but her duettist. There in the woods around her home, a nightingale sang, seemingly prompted by the sound of the cello. It occurred on a number of successive evenings, and was predictable enough for the BBC to set up one of its first live outside broadcasts to capture the happening. Harrison later described the event in her autobiography:

> Suddenly, at about a quarter to eleven on the night of the 19th May, 1924, the nightingale burst into song as I continued to play. His voice seemed to come from the Heavens. I think he liked the *Chant Hindou* best for he blended with it so perfectly. I shall never forget his voice that night, his trills, nor the way he followed the 'cello so blissfully. It was a miracle to have caught his song and to know that it was going, with the 'cello, to the ends of the earth.[2]

The whole thing had started some nights before, when Harrison was rehearsing at home. The bird started singing, as if responding to her playing. Beatrice contacted John Reith, general manager of the fledgling BBC, and it was agreed that this was a moment that could put this new medium firmly on the map. Here was an opportunity for a radio revolution: a 'live' outside broadcast. The

---

[2] B. Harrison, *The Cello and the Nightingales* (London: John Murray, 1985), p. 132.

date of the 19 May was agreed, and the event was trailed with plenty of advance publicity. Richard Mabey has explained what happened next:

> Two van-loads of equipment and a large battalion of engineers arrived ... and set up the operation in the garden. The microphone was placed as close as possible to the bird's singing post, and the amplifiers were stacked in the summerhouse ... The plan was to wait until the bird was in full song and then break into the Savoy Orpheans' Saturday night dance programme.³

The weather was perfect, the moon was full and the air was still and warm. Yet despite many attempts, and various musical offerings from Beatrice, the bird would not sing. The clock ticked on, Beatrice played, but her cello sang alone.

> Then, just after 10.45, twenty minutes before the station was due to go off the air, the bird began. An excited continuity announcer broke into the Orpheans, and for the next fifteen minutes the BBC's audience listened entranced to the historic duet.⁴

The duet between human and bird gained almost instant fame, becoming a universal phenomenon, to the extent that some years later, when King George V met Harrison, his first words were: 'Nightingales, nightingales! You have done what I have not yet been able to do. You have encircled the globe'.⁵ The woodland around the house was a haven for birds, and prompted by the success of the initial broadcast, subsequently other birdsong transmissions and a commercial recording were made in Beatrice Harrison's garden, all achieving great success and sales. For nearly one hundred years, the Harrison/nightingale broadcasts were acclaimed as one of the iconic moments in radio history. Yet all was not quite as it seemed. In 2009, the author Jeremy Mynott raised some doubts as to the authenticity of the song in the 1924 recording; there was something about the sound of the bird that struck him and other experts as curious. Opinion was divided, but Mynott presented a persuasive piece of evidence in the form of a letter from Ted Pittman from Kent, the grandson of one Dame Maude Gould, a bird impersonator – a siffleur – who worked under the name, 'Madame Saberon'. His account stated that the BBC had booked Maude to be present on the night of the broadcast, as a backup in case the nightingale did not sing, and it was in fact

---

³ R. Mabey, *Whistling in the Dark: In Pursuit of the Nightingale* (London: Sinclair-Stevenson, 1993), p. 101.
⁴ Ibid.
⁵ B. Harrison, *The Cello and the Nightingales*. p. 132.

her voice, and not the bird's, that enthralled the audience. 'The trampling around of all the technical staff and all the heavy equipment scared any birds off and the recording is actually that of Maude Gould whistling to Ms Harrison's playing'.[6] A further complication lies in the fact that at the time of the broadcast, the BBC did not have the facilities to record outside broadcasts. Disc recording was still a complex and time-consuming process, and the question arises as to whether the recording that has been passed down to us represents that very first event, or actually a subsequent 'performance' made by a commercial record company? Whatever the reality, the effect at the time was colossal, and a tradition of cello/nightingale broadcasts followed annually for a number of years, all of which were, by all accounts genuine.

It also needs to be put into the context of the nature of 'recorded truth'. In early recordings, there are a number of precedents for reproducing spoken texts in other than the original voice; it was the *content* that mattered, the spirit of the thing, almost as much as authenticity. Gladstone's voice was first recorded by Edison, but due to technical reproduction difficulties, his speech was recreated by actors on subsequent recordings. Likewise, years later, some of Winston Churchill's most famous wartime speeches were recreated for overseas audiences by an actor. Within these parameters, therefore, Madame Saberon was being morally true to the spirit of the heart of the matter in her recorded text, quoting the nightingale's song. Thomas Hardy's description of woodland sound exists, and remains a record of a place and time, while Madame Saberon's nightingale facsimile created another kind of text that formed a concept in the minds of those who heard it, establishing a seminal broadcasting moment. Thus in the telling, and in their own ways, they both remain as stories in which memory and imagination perpetuate an event that continues to exist in a different reality. What persists above all is the story itself, which has gone into broadcasting history and lore. If the later version of events is itself to be trusted, deception of the audience did not, apparently, come into the equation at the time, and the tale has been told and retold in numerous books (including my own), on media and the early years of radio, and continues to capture the imagination of new generations coming to it for the first time. The fact is that by now many more people have read *about* the Surrey nightingale duet than have actually *heard* the recording itself. It is almost as though we don't need to witness it to be captivated

---

[6] T. Pittman. Quoted in J. Mynott, *Birdscapes: Birds in Our Imagination and Experience* (Princeton: Princeton University Press, 2009), p. 313.

by the idea. It is recorded in words, and knowing that it happened, and can be told, is enough, and conveys its own reality to the mind, just as Hardy's wood exists in the imagination through his written text.

Early in broadcasting history as the nightingale outside broadcasts were, they were not the first attempts at holding on to the sounds of avian life. In 1889, the eight-year-old Ludwig Koch made what has been claimed to be the first surviving sound recording of a bird: an Indian Shama bird. Koch is a key player in this story, and we shall devote time and space to him later. Between his pioneering wax cylinder recording and Beatrice Harrison's broadcast encounter with the Surrey nightingales, the preservation of sound as an artefact developed at a remarkable rate. In the same year as Koch recorded his Shama Bird, the German-American inventor Emil Berliner was working on what became known as the gramophone. More importantly, he was also developing a method of recording that involved a stylus on a disc rather than a cylinder. From 1896, this disc-cutting device was powered by an electric motor and could therefore run at a constant speed. A 7-inch diameter disc could supply two minutes of recording time, and in 1901, larger discs increased the available duration first to three minutes and then in 1903 to four minutes, from a 12-inch disc. Koch, predictably, was among the first to avail himself of this technology, mostly within domestic and garden environments.

Meantime, in England, the photographer and writer Cherry Kearton managed to record a few notes of a nightingale and a song thrush, using the old cylinder technology. Cherry and his brother Richard were naturalists who were early specialists in wildlife photography, and in 1895 they had published the first natural history book to be entirely illustrated with photographs. In 1910, Karl Reich from Bremen in Germany released the first commercial record of a bird – a captive nightingale – a disc which contained on the 'b' side the sound of a thrush, thus replicating the Kearton cylinder recording on new technology, via the Victor Talking Machine Company in Camden, New Jersey, and making it available to a wider listening public.

The naturalist and wildlife broadcaster Eric Simms identified the first known recordings of wild birds in America and Africa as being much later, in 1929, and it was in Australia in 1931, on 27 June, that the song of a lyrebird was committed to disc. There is also a reference in the account of the 16th Congress of the American Ornithologists' Union as early as 1898 which refers to the playing of 'a graphophone demonstration of a Brown Thrasher's song'.[7] In the early

---

[7] E. Simms, *Wildlife Sounds and Their Recording* (London: Paul Elek, 1979), p. 3.

recordings, the use of caged birds was often considered expedient due to the limitations of the technology; pre-electric mechanical and acoustical horn-based machines were limited in their top and bottom frequencies, and had to be placed close to the subject in order to capture any meaningful sound at all. At the Paris Exhibition of 1900, a wire recorder – the Telegraphone – was demonstrated, 'two years after Poulson magnetized a steel wire of 0.5 to 1mm diameter from a piano with alternating fields of audio frequencies and, by running the wire alongside a coil, reproduced the alternation with the help of the magnetism left, as a sound in a telephone'.[8]

This then, briefly, is a hint of the prehistory of the coming of recording and its application in the preservation of natural sounds. Up until this period of technical activity, the word 'text' had been applied to words on paper; things would now never be the same again.

It was not, however, until the arrival in 1925 of the electrical broadcasting microphone that real progress was made in terms of quality, enabling both music and wildlife to be caught in a new clarity that did at least some justice to the original. This was of course after the first Harrison broadcasts; nonetheless, a duet between a cello and a bird caught the imagination, and this and subsequent broadcasts and recordings opened a door, through the dramatization of the event. Until that point, most of the people who listened to the duets would not have heard the actual sound of a nightingale in the wild. However, they might well have read poems or accounts of its song, since the nightingale is probably the most documented and celebrated voice in the avian world. These people would have gained an *idea* of the sound, conveyed through words, with varying degrees of success, dependent on the skill of the writer and the acuteness of their perception. Today technology allows us to listen to birdsong wherever we are; we can be familiar with sounds from far-flung countries we may never visit in person, and not only become experts on the nuances of avian signals, but share them in the comfort of our own home also. Yet before the advent of recorded sound, the written word was the only real tool for bringing the experience of birdsong to the masses, beyond being present as a witness to the

---

[8] Ibid., p. 4. Valdemar Poulson was a Danish engineer whose contributions to the development of early radio technology were considerable. The Telegraphone was his invention, and in 1903 he went on to develop the Poulson Arc transmitter, the first continuous wave transmitter, which was later used in some of the first radio stations up until the early 1920s.

actual event. So how was it described during that time, and how very different was it from today's first-hand data-capture?

The thirteenth-century rota, or round, 'Sumer is icumen in', sometimes known as 'The Cuckoo Song', is one of the best-remembered and most frequently quoted renditions of the sound of a bird as a representative of a season beginning, and may have been the first poem I actually read, coming as it did on the opening page of my father's 1931 edition of *The Oxford Book of English Verse*:

> Sumer is icumen in,
> > Lhude sing cuccu!
> Groweth sed, and bloweth med,
> > And springth the wude nu –
> > > Sing cuccu!⁹

To my childhood ear, this was pure sound, and the onomatopoeia in the repetition of the word 'cuccu', itself a name based on the voice of the bird, evoked then as now, an unequivocal representation of the noise of spring:

> Awe bleteth after lomb,
> > Lhouth after calve cu;
> Bulluc sterteth, bucke verteth,
> > Murie sing cuccu.¹⁰

It is as though one living thing sends a cue to the rest of the natural world to awaken; the sound of the bird rings familiar even through the Wessex dialect of Middle English, itself almost a set of abstract silent musical notes, to the modern ear. We do not know the author, but some sources suggest it was William de Wycombe. Whatever, it is acknowledged in its musical form to be the earliest musical composition featuring six-part polyphony, but even as a spoken poem, it is pure sound, and makes its own music even before it is sung. In the final verse, it gives itself over to a tumult of ecstasy, poet and bird encouraging one another's song.

> Cuccu, cuccu, well singes thu, cuccu:
> > Ne swike thu naver nu;
> Sing cuccu, nu, sing cuccu,
> > Sing cuccu, sing cuccu, nu!¹¹

---

[9] A. Quiller-Couch (ed.), *The Oxford Book of English Verse* (Oxford: Clarendon Press, 1931), p. 1.
[10] Ibid.
[11] Ibid.

The sound recordist, Chris Watson, recording on Holy Island, Northumberland, in the twenty-first century, sought to evoke a seventh-century soundscape such as the monks who created the Lindisfarne Gospels might have heard as they worked in the fields, or laboured over the great manuscript. It was a ninth century Irish scribe who memorialized the cuckoo there in verse some three hundred years before that same song caught the attention of the spring song's author:

> A clear-voiced cuckoo sings to me (goodly utterance),
> In a gray cloak from bush-fortress.
> The lord is indeed good to me:
> Well do I write beneath a forest of woodland.[12]

This raises a key argument of this book. A sound recording is what it is: a record, and the definition of a record is both as a noun and as a verb. It is a thing constituting a piece of evidence about the past, initially, historically and primarily an account kept in some permanent form. It may be written as text, or latterly, in the form of signals on wax, shellac, tape or in digital form; it remains an evidential artefact that does no more than register proof that an event happened. Thus, the verb, *to record* articulates the intention to enable the existence of such evidence. When we seek to describe, not only the sound, but the *effect* of the sound, we find descriptive creative literature to be our refuge. 'A clear-voiced cuckoo sings' is a record, but 'A clear-voiced cuckoo sings to me' is an acknowledgement that I was there to hear it and that I am a beneficiary of the sound (something of which the cuckoo is unaware, and about which the bird does not care.) Further, 'A clear-voice cuckoo sings to me (goodly utterance)' is a value judgement that evokes feeling. The scribe's poem thus tells us not only of the sound, but of the way it changes his environment, and himself.

It is so often the cuckoo that gains representation, either in words or in music. This simple two note song is both instantly replicable, and easily simulated. Because of this, it is the first bird to be recognized, its song carries, usurping the air as the bird itself usurps the nest. It is thus a shorthand for many composers seeking to evoke a landscape. Beethoven (1770–1827) in his *'Pastoral' Symphony* 'records' it, Gustav Mahler (1860–1911) places it near the start of his *First*

---

[12] C. Watson, *In St Cuthbert's Time: The Sounds of Lindisfarne and the Gospels* (Newcastle-upon-Tyne: Touch Music). CD insert booklet, note by Dr Fiona Gameson, St Cuthbert's Society, Durham. CD Number Touch to: 89, 2019.

Symphony, while the English composer Frederick Delius (1862–1935) gives us a whole tone poem, *On Hearing the First Cuckoo in Spring*, motivated by the same emotion that fired the Irish scribe to write in praise of the bird. The music not only mimics the song, it also describes the mood of the moment, a microcosm of the burgeoning season surrounding it. For many composers, birdsong is a decoration rather than the main subject, although the Italian Ottorino Respighi (1879–1936) placed avian life centre stage in his 1928 five-movement work based on the seventeenth and eighteenth century music, *The Birds*, in which he attempted to transcribe birdsong into notation, and even include bird actions, such as scratching feet or fluttering wings. At various points in the piece, we hear the dove, the hen, the nightingale and the ubiquitous cuckoo. Olivier Messiaen (1908–92) went further in 1952 when he was invited to provide a test piece for flautists at the Paris Conservatoire; his response was to compose *Le merle noir* for flute and piano, based entirely on the song of the blackbird. He had long been fascinated by birdsong and had incorporated it into a number of his earlier works, but the idea of transcription took hold for him with the Paris work, and the following year, 1953, other pieces began to appear, such as his work for orchestra, *Réveil des oiseaux*, evoking birdsong heard in the Jura Mountains region of France, culminating in the famous *Catalogue d'oiseaux* of 1958, which gives a sense, not only of the bird, but also the mood context of its location. His massive *Turangalîla-Symphonie* of 1949 takes birdsong into a richly imaginative and complex tapestry, and shows Messiaen to be, of all composers up to his time, one of the most enthusiastic and knowledgeable ornithologists of all composers. Subsequently, a late work by the English composer Jonathan Harvey (1939–2012), *Bird Concerto with Piano Song*, combined recordings of real birdsong with electronics, orchestral music and a pianistic commentary, of which Harvey wrote: 'Real birdsong was to be stretched seamlessly all the way to human proportions – resulting in giant birds – so that a contact between worlds is made'.[13] The work is haunting and moving, although its blend of real birdsong with simulation drew criticism from some ornithological quarters. The clustering sound evokes flashes of light, perhaps echoing the morning air in California, where the *Concerto* had its genesis. At the work's root is the idea of the song, as Arnold Whittall wrote: 'The overriding aesthetic quality of the piece throughout it, that of "playing with the idea of song", the basic contrast

---

[13] J. Harvey, Programme note for CD insert booklet. In *Bird Concerto with Piano Song* (London: NMC, 2011). Records NMC D177.

between organisms which sing and instruments which can (only) play [fuelling] a transformational drama'.[14]

Composers may incorporate the song of nature either through specifics or in more general ways. Between 1895 and 1897, Maurice Ravel (1875-1937), wrote a two-movement work for twin pianos called *Sites Auriculaires* (broadly translated as 'places which might be sensed by the ear'). The first of these was *Habanera*, inspired by his Basque mother's stories of her young life in Madrid, conjured for her thereafter by this song form. The second, *Entre Cloches* ('Between Bells'), evoked through the counterpoint of bell sounds in the piece, the sounds of various Parisian church bells tolling together at noon. If there is any birdsong here, we may imagine it to be drowned out by the clamour of the bells or replaced by the flutter of startled and complaining wings. For many other composers, however, birdsong may be the central subject evoking Place, or the audible part of a visual landscape, more than décor, the subject itself and sound that gives meaning to the place in which it is heard. Tori Takemitsu (1930-96), in his work, *A Flock Descends into the Pentagonal Gardens*, invites us to approach a garden as from a distance, with the oboe as the principal voice, and the circling of birds, pointing to the fact that when we think of the sound of birds, we should never forget the fluttering and beating of wings. As children, we became familiar with orchestral birds: Serge Prokofiev's *Peter and the Wolf* evokes woodland and its creatures, Leopold Mozart's (1719-87) *Toy Symphony* gives us particular birds in the sound of the cuckoo, the quail and the nightingale, while Haydn's (1732-1809) *Symphony no. 83 (The Hen)* is self-explanatory, perhaps gaining inspiration from other works such as the early seventeenth century Carlo Farina (1600-39) from the Court of Dresden, who peopled his 1627 work, *Capriccio stravagante* with cackling hens and barking dogs. Antonio Vivaldi (1678-1741) too, in his *Four Seasons* gives us birds, dogs and insects as part of the musical world. Famous among musical representations of the animal kingdom is Camille Saint-Saëns's (1835-1921) *Carnival of the Animals*, with its aviary, woodland birds and swan, this last being an expression of mood and movement rather than song, as with Sibelius's (1865-1957) dark *Swan of Tuonela*. The Finnish composer, Einojuhani Rautavaara (1928-2016), in his *Cantus Arcticus, Concerto for Birds and Orchestra*, explored similar artistic if not geographic territory in 1972, employing recordings of bird sounds

---

[14] Ibid.

made in the Arctic Circle and marshlands of Leminka. Orchestra and waterfowl seemingly interact, and we hear the calls of waders, shore larks and migrating swans both individually and collectively, and benefitting from the technology that enables as it were, 'wide-angle' and 'close-up' auditory images to be gathered. All these compositions are texts, written forms translated back into sounds through orchestras or other combinations of musicians, and while they *are* textual interpretations of a sonic event experienced in the natural world, they are not the main subject of this book. Much has – and will – be written about music as a representation of, or emotional response to natural sound, but here I want to focus on the *word* as sound and thought-transmitter. Avian life is at the root of music, but so often we find composers stepping back from the challenge of matching it, using the sounds as stylised jumping-off points for expression, because, when all is said and done, the sounds of the natural world are sonically inimitable.

Even when the first recordings of birdsong were made, miraculous as they seemed, they were technically rough and indistinct, reminders, souvenirs, suggestions of the real things. Little wonder that the idealized sound of the natural world was prized in words and in music. George Meredith wrote his poem, 'The Lark Ascending' in 1881. It was after the First World War that Ralph Vaughan Williams took it as the inspiration for his soaring single movement work for violin and orchestra, when, in the wake of four years of carnage, the metaphor of its ecstatic innocent flight into invisibility must have been deeply poignant and meaningful. The piece is now a popular favourite amongst concert audiences, but how many of those rapt listeners have been drawn to seek out the real thing? Meredith's poem expresses the sound perfectly:

> He rises and begins to round,
> He drops the silver chain of sound,
> Of many links without a break,
> In chirrup, whistle, slur and shake,
> All intervolved and spreading wide,
> Like water-dimples down a tide
> Where ripple ripple overcurls
> And eddy into eddy whirls[15]

---

[15] G. Meredith, *Selected Poems* (Boston, MA: Elibron Classics, 2005), pp. 5–6.

Had we been with Meredith on the downs where he witnessed the bird, our ear and eye would have followed his, up, up into the air, striving for a last glimpse, a final hint of sound:

> Till lost on his aerial rings
> In light, and then the fancy sings.[16]

It is indeed the fancy that sings when we read the words, and that is the important thing, because Vaughan Williams gives us his sound version of the bird, evoking a mental picture based on what he heard in his mind, informed by the poem and his own memory, while Meredith, through suggestion, allows us to have our own. There is at the heart of all this, a closeness yet simultaneously a remoteness that highlights the strangeness of our relationship with the animal kingdom. We share our world with creatures that pass us by; they see us and sing from our trees, and yet there is a point beyond which we cannot go, a line of understanding we can never cross, and which we can only interpret and mimic. All of which leads us to a question: now we have the means to perfectly record the songs of birds, why should we seek to record them in the written word? Perhaps part of the answer is to be found in the shift of mindset that pandemic lockdowns, wars and the displacement of communities provoke, throwing us back into a state in which a sense of place is felt acutely, and our awareness of immediate surroundings is in turn thrown into sharp relief. Another answer to the question may be that while a recording gives us a sense of direct access to the event, and while it can have an equally direct effect on our emotions and senses, it is through 'hearing' or reading the impressions of a third party that a discussion between the intellect and the senses is initiated. Thirdly, of course, birdsong is inspirational; it opens the capacity for song in us, however we may interpret the idea of that, and we are moved to express our own personal responses. There is yet another reason, an important one which grows from that, and in a sense, it underpins all the others. It is to do with our subjective response: in other words, it has more to do with our state of mind at the time of listening, our personal circumstances and our mental response and attitude. The bird's call may affect us, dependent upon our feelings at the moment of hearing it, while the recording remains the same, because a microphone cannot feel emotion, or think. Tom A. Garner has summarized this variance while listening to birdsong, and his words equally apply to 'live' and recorded location-based sound:

[16] Ibid., p. 8.

During exposure to the birdsong soundwave, the listener may experience multiple distinct perceptions that merge together as a singular dynamic experience. For example, our listener's immediate affective response to the birdsong may be positive as it prompts the recall and imaginative re-experiencing of a happy memory, say a romantic encounter. This influences the listener's behavioural response as they adjust position to better hear the song. As the soundwave continues, the realisation that the relationship (with which the happy memory is associated) is no more, dramatically changes the emotional context of the birdsong and the listener turns away from the sound source in an attempt to prevent any further painful thoughts. Here, both endosonic [psychological] and exonsonic [physical] elements are changing throughout what the listener would understand to be a single, fluid experience.[17]

Most written accounts of birdsong and the wider natural world in acoustic terms are created 'in tranquillity', with hindsight and with the benefit of reflection, and thus the potential for the recollection of emotion is sometimes subverted by other feelings. Memory is a key part of the nature of human interaction with all sound, but the highly charged sensory experiences of the natural world heighten the response.

In the end, it is all text, whether it be on parchment, paper, stone, disc, tape, digital memory card or carried inside the mind itself. So we go on recording, and we go on writing about the experience. The writers we shall consider later – Gilbert White, John Clare, William Wordsworth, Henry Thoreau, Richard Jefferies and the other poets of the natural world writing before the advent of recording technology – sought to convey a sense of what it was like to be *there*, often to an audience of readers for whom the experience itself was denied, or at least distant. In doing so, they also preserved the moment through the silent sound held in words on a page. When we visit White's home village of Selborne, Thoreau's Walden Pond, or the Wiltshire Downs familiar to Jefferies, we hear not only the world around us now in our time, but in the printed pages they left us as witnesses to their time in the same place. Through their words, we can imagine how it was for them, and how it made them feel. Perhaps because of the changes to the world, or perhaps despite them, there has also been a growing richness in modern writing about the natural world, and an awareness of something so

---

[17] T. A. Garner, 'Bridging the Other-Real' in M. Grimshaw-Aagaard, M. Walther-Hansen and M. Knakkergaard (eds), *The Oxford Handbook of Sound and Imagination, Volume 1* (Oxford: Oxford University Press, 2019), p. 797.

nearly lost, much as there was in some of the words written around the first years of the twentieth century. Are we becoming better listeners? In his remarkable book, *Diary of a Young Naturalist*, Dara McNulty writes of a visit to Rathlin Island, off the coast of County Antrim in Northern Ireland:

> After dinner, song bursts from every corner of the sky and we stop to listen in the twilight. Isolating each and every melody, I feel suddenly rooted. Skylark spirals. Blackbird harmonies. Bubbling meadow pipits. The winnowing wings of snipe. And always the sound of seabirds. We are in the other world. No cars. No people. Just wildlife and the magnificence of nature.[18]

Through his diary, McNulty describes a year in his life as a passionate lover of the natural world, and as an autistic teenager for whom THIS is the real world, first and foremost; in doing so, he finds words to describe it in such a way that he peels back a screen, and removes a filter, to convey the moment:

> A solitary gannet scythes the sky, and its cantering cries synchronise with my heartbeat ... and as the rain falls, I feel the warmth of its lamenting calls.[19]

If we listen harder, we may care more. There are species of birds and other wildlife that are no longer audible, because they are no longer there. The natural world of England does not sound the same as it did in Gilbert White's time, not only due to the decline in species caused by climate change, but to one of the major contributory factors to that decline, the roar of aircraft overhead, and traffic, so often sounding from beyond the hill, where the motorway rumbles, or where the town or city roars. The best writers of impressions communicate as though today could be the last in history. Nothing is taken for granted, nor should it be. We listen to their words because it is another person talking to us, and telling us, 'This is what it was like when I was there'. No two moments are the same, sound is linear, it passes us and is gone. What greater metaphor can there be for the urgency and importance of recording – in whatever form we care to use the word.

This journey will end in an exploration of some of the contemporary written accounts of sound in the natural world, evidence that there will always be a human need to respond to beauty. Some of the most transporting sounds I know have come to me through my headphones, and better still, standing in a wood or on a hilltop, listening across a valley to the perspective of sounds from different

---

[18] D. McNulty, *Diary of a Young Naturalist* (Beaminster: Little Toller Books, 2020), p. 25.
[19] Ibid., pp. 29–30.

directions and distances, the call and response of birdsong; at the same time, some of the greatest expressions of the human spirit are there to be had through the opening of a book and the turning of a page, as we read how such experiences affects or affected others. Take the opening of Richard Mabey's book about the Suffolk nightingale, *The Barley Bird*, written in 2010:

> It's early May, a nightingale moon. I'm perched in a narrow lane above the Stour valley in Suffolk, listening to the birds ... It's my first nightingale of the year, and a coloratura bird into the bargain. It has a clipped, Latin style, full of deft phrases which are turned this way and that, and drawn into short, fading tremolos ... The sound is astonishingly pure and penetrating, broken with teasing, theatrical silences.[20]

It takes a unique writer with a unique ear to convey an experience with such precision and beauty, but it takes a nightingale to sing it. Above all, it is precisely passages like that, in which I can 'hear' a voice defining a sound, that demonstrates the point I want to explore. Yet this journey cannot hope to be by any means comprehensive; the most it can aim to do is to celebrate the human interaction with nature in a relatively small handful of textual representations of its sounds, calls and songs through a few selected examples that inevitably must reflect the personal knowledge and choices of this author. In so doing, its intention is that of advocacy towards an informed and active sense of the auditory, and a consciousness that listening is an art to be developed and cherished through the partnership of hearing and mindfulness.

An early step on the journey will take us back to radio, and the work of Ludwig Koch, whom we have already mentioned. Koch belongs to three literatures: technical recording, the printed word and the natural world. That natural world, the sound of it, IS itself a text, printed on air and in the earth, so when we record it, or write about it, we are really only transcribing what is already there, in order to find meaning in it. Subsequent chapters will visit some of the random, almost miscellaneous, and incidental moments when a writer has heard something and sought a way to express that event. There are voices of troubadours, Chaucer and Shakespeare to explore, the thread of science informed by poetry in the work of John Ray and Francis Willughby, Gilbert White and Izaak Walton, the writings of Wordsworth, Keats, Hopkins

---

[20] R. Mabey, *The Barley Bird: Notes of the Suffolk Nightingale* (Woodbridge: Full Circle Editions, 2010), p. 15.

and others for whom birdsong opened doors into new and strangely interior worlds, Richard Jefferies, Henry Thoreau, and Edward Thomas, the last, writing just on the cusp of technical recording and vast societal and industrial change. Across the centuries, it is all to do with listening through the creative impulse, seeking to *understand* what we hear, and what it means.

Despite all the modern technical facility, the ability to preserve, playback, alter speed, volume, and tone to delve into the minutiae of auditory detail, we are still moved to respond in the same way as that ninth century Irish monk: 'Well do I write beneath a forest of woodland'. We do it partly because of the wonder of it, because it is too amazing not to remark on, and partly because, in listening, we are confronted with a mystery. The vocal organ of most birds is the syrinx, a beautiful word, with a Greek origin that means 'pan pipes'. It is significant that Claude Debussy (1862–1918) wrote an exquisite and mysterious piece for solo flute with that title. The anatomy of birds tells us how much the shape and design of this organ varies, but it is interesting that the beauty of a song does not necessarily arise from a particularly complex syrinx. In their book, *Bird*, Lois and Louis Darling spend some time examining the shape and structure of a number of bird larynxes and syrinxes in order to reveal the mechanism that makes the song, only to conclude it to be 'a subtle mechanism whose function would be as lovely to the eye of any poet, should they but see it, as the sound of the song to their ear'.[21] There is then the connection between the physical and the emotional/spiritual to consider, something that, as Henry David Thoreau put it, '[cannot] be represented on canvas or in marble only, but [is] carved out of the breath of life itself'.[22] That has been the almost eternal challenge to creative and descriptive writing and it is the purpose this book has set itself to explore.

The fact is, we need both words *and* actuality, now more than ever. In both forms, there is an urgent need to preserve the witness of what is *our* experience, for future generations. They are records in every sense; we document what is, so that it can exist beyond us. Species die out, the world changes, grows sparser. The specific sounds of nature are like jazz, changing every time we hear them, because the surrounding ambience – the backing track that is the accompaniment, the mixtape, an integral part of the music itself – keeps coming up with an infinite range of arrangements; so no one performance is definitive, while every single hearing is unique. Hardy's wood in Dorset doesn't sound the same as it would

---

[21] L. Darling and L. Darling, *Bird* (London: Methuen, 1963), p. 173.
[22] H. D. Thoreau, *Walden* (London: Penguin Books, 2016), p. 96.

have done when he was there as a boy in the middle of the nineteenth century, nor even when I was standing above his cottage as the Falklands War began. His words on the page are all I have, to preserve a sense of how he felt in that space, while mine, reproduced here as a memory, are all I can offer you in terms of what it was like to be there in my day. To sit with a text is to gain a context for emotion triggered by an event, while to replay the sound of the event itself is to allow, at least in part, the listener to enter a new imaginative emotional moment of their own.

2

Recording an essence: Hearing, reading and the imagination

Playing through some field recordings with an ecologist friend, we were discussing the minutiae of sound, the tiny happenings that are constantly going on around us to which we pay little or no heed, of which we are mostly completely unaware. As an example, he took out his smart phone, saying 'listen to this'. It was a fifteen second clip of apparently random ambience, recorded in a wood. 'What do you hear?' he said. I listened, then listened again. Quite general sound, some nearby birdsong and some rustling and clicking. Nothing particular to remark upon. He drew my attention to the tiny 'clicks' which at first I'd taken for twigs snapping underfoot. 'That', he said, 'is the sound of pine cones opening.' It was after a spell of heavy rain, we were enjoying some surprisingly hot weather, the sort of day that can sometimes come towards the end of March. This was the sound of a response to that temperature change. So small, so apparently insignificant, and yet it was *there*, it had happened; easily missed, but as much a part of the turning of the seasons as all the vivid and spectacular budding, leafing and blossoming that was about to erupt across the land.

In the spring of 2020, countries closed their borders, and neighbourhoods shut their doors as the Coronavirus struck. Traffic almost vanished from roads, city centres grew quiet, offices and public places were shut and air travel was suspended. It was a palpable sonic change to life; some even remembered the moment in 2001 when the flight lanes emptied in the immediate wake of 9/11 attacks. It was a time when we became aware of sound and silence, as though a veil had been drawn back. As it happened, in England, the lockdown coincided with the most exquisite spring season on record. As pollution levels – both auditory and atmospheric – fell, the natural world responded. You could almost hear the leaves opening. Even many who were not given to active listening noticed it. The light seemed to clear before our eyes, and it was as though the air carried a new purity of sound. Partly as a result of this, there was a growth in interest in field recording; some sound professionals who had previously used their equipment to record voice and journalistic content, became fascinated by capturing ambience, the sound of the world happening before their ears, making podcasts and websites out of the audio environment, the signals that had been in some cases there all the time, but covered in layers, as an ancient mosaic floor is gradually covered by coatings of silt and soil over time. We turned to our digital recorders, aware that this would not last long, this return to something that might just be a clue to how the world once sounded, and could sound. We tried to hold time in our hands, to freeze it and preserve it.

The ability to do this had been gradually refined over more than a hundred years, from the late nineteenth century's faint ghosts coming to us through the fog of hiss and crackle, right up to cutting-edge twenty-first-century sound technology, the digital files in palm-held machines that we could carry anywhere, ready to catch the instant, like a camera. There was a sense that the technology had become our ears, so the fleeting moment that came out of silence, passed us and moved on, was in some mysterious way, saved from oblivion. The first wonder was that we could do it at all; but that did not last long. Very quickly one of the first and most enduring realisations that comes with the ability to preserve, dawned on us: what IS it we have preserved? And how much is it a true representation of the reality to which it was a witness?

We must turn, as so often, to original definitions. The word, 'Record', like its close companion through the twentieth century, 'broadcast', is both a noun and a verb, and comes from an earlier time. 'To broadcast' in the media sense, was a word used for the dissemination of a signal, but its origin was agricultural, a word dating back to one meaning: 'Of seed, etc, scattered over the whole surface.'[1] Both words were appropriated by new technology; 'to record', as understood by someone working in media, means to preserve in a format that lends itself to playback. From early foil and wax, to discs, wire, tape in various forms, on to digital files, 'recording' has always served one purpose, that of catching and holding an incident in time. 'A record' as a noun is the product of that process. Yet its origins are with the written word: 'The fact or attribute of being, or of having been, committed to writing as authentic evidence ... The fact or condition of being preserved as knowledge, by being put into writing, knowledge or information preserved or handed down in this way.'[2] This brings

---

[1] *The Shorter English Dictionary*, vol. 1 (Oxford: Oxford University Press, 1973), p. 240.
[2] Ibid., vol. 2, p. 1766.

home the profundity of the human need to save experience from oblivion, and applies to text in all its forms: written, illustrative, visual and sonic. The nuances become simpler and even more moving when we consider the verb, 'to record': 'To bring to remembrance ... to relate in writing ... to get by heart, to go over in one's mind' and then there is this: 'Of birds [rarely of persons]: to practise or sing a tune in an undertone, to go over it quietly or silently'. The compiler of my edition of the Shorter Oxford English Dictionary (*OED*) is even moved to quote the words of Charles Darwin to make the point: 'The young males continue practising, or, as the bird-catchers say, recording, for ten or eleven months.'[3]

Can the original essence be held like this, preserved exactly? That is to say, can the *experience* of the actual moment be transcribed to the extent of replication in every detail of the reality of the confluence of place, temperature, mood and perspective? When does memory become part of the process? For it is not only about the act of recording; even more, it is to do with the art of listening, and *mentally, emotionally* recording the many micro-incidents that go to make up the world as it happens around us, that matters most as we move second by second away from the current moment. All recording, be it through word or technology, is to do with transmission of experience through time and space, from one place to another. Thus, the sound of a waterfall in North Wales may be 'heard' in the Sahara Desert and the song of the last bird of a species that died a century ago may be played back and experienced today. It may be as a technically recorded sound, or as a set of words, phrases and sentences – themselves sounds – set down in a certain order. Either way, there is an instinct for preservation, to convey to the imagination, the spirit of the sound and the creature that once made it.

It was during that strange and terrible time in early 2020 that we somehow, even fleetingly, grasped an understanding of how much listening was part of our very existence, and how we possessed the ability to be ourselves, recorders through memory if we chose to, if we could but focus and concentrate that facility within ourselves. This was sonic history, and we were witnesses to it. Some made soundscapes. Others tried to hold it in words. In their book, *The Consolation of Nature: Spring in the Time of Coronavirus*, writers Michael McCarthy, Jeremy Mynott and Peter Marren did the latter, by sharing the task through journals that

---

[3] C. Darwin, quoted in ibid.

would act as a mnemonic for the rest of us as that strange hiatus receded into the sepia of time. Turning its pages opens those days. What their writings do that the field recordings of the time cases do not, is preserve human responses to the experience as diary entries. For the casual sound recordist, there is a risk that the machine is left to do the listening, as a snapshot taken quickly on a phone is charged to do the looking, and that because we know the sound or image is safe, we turn off part of our attention and relax our memory. To listen and observe, and at the same time observe the process itself, and its effect upon us, is to take a recording somewhere else. Not only do we retain the sound, and a sense of Place more strongly, the recollection may lead to new paths of discovery, as we delve into meaning, culture and history. Here is Jeremy Mynott, listening to chiff-chaffs near his home in the village of Little Thurlow in west Suffolk, during the week of 21–25 March 2020:

> There must be at least four of them chanting *chiff-chaff* in the copses here today. I like some of the old country names that perhaps capture their disyllabic song better: *chip-chop*, *chit-chat* or *siff-saff* (a Welsh version). Other European countries hear it differently: *zilp-zalp* (German) and *tjift-tjaf* (Dutch) are both nice, but the specific name *collybita* is perhaps more imaginative: it means 'money-changer', that is 'coin clinker'. I realise that the clarity of these and other birdsongs I'm hearing all round me is intensified by the lack of any traffic or aircraft noise. This is so rare in today's mechanised world that the silence has the force of a new and positive presence, a medium in which the natural world can more fully express itself.[4]

All this came at a time in which we could also reflect on the urgency of planet preservation, as the reality of climate change came home in the most tangible and visceral ways. By paying attention to the sounds in this newly purified air, we were listening to a responsibility to save it for future ears. That same week in March, 2020, I was writing in my diary from a very different location, in a suburb of south Liverpool, on a normally busy arterial road, less than five miles from John Lennon Airport:

> The avenue is still and traffic-free this morning at eight o'clock, when normally it would be a river of cars and buses heading into the city. The airport runways are silent: we hear nothing but birdsong at this time. And the cherry blossom

---

[4] M. McCarthy, J. Mynott and P. Marren, *The Consolation of Nature: Spring in the Time of Coronavirus* (London: Hodder Studio, 2020), p. 17.

trees that line our part of the road, all along the central reservation, are shining with an almost blinding whiteness in the sunlight. Birdsong and blossoms; there may be fear and uncertainty across the world, but nature doesn't appear to have got the memo.[5]

A week later, I was in my back garden with a Tascam digital recorder and noted:

> At around eleven this morning I captured, within less than a minute, the songs of Greenfinch, Chaffinch, Bullfinch, Goldfinch, Robin and Willow Warbler. Such pure sound in light made noticeably clear by the reduction of air and ground traffic.[6]

I notice now, the word 'captured'; it's a word we often use, especially with sound or photography: 'Great capture' we say, but what has been captured, and from what? An essence. Something has been caught in the nick of time, before everything changed and moved forward. Listening to these songs, and playing them back that evening, it was hard to remember that we were in the midst of an existential crisis, and that while I listened to the life burgeoning in my back garden, along the streets at the front of my house, there was a strange, unfamiliar silence, the city empty, people afraid to go out, shops closed; people dying in their thousands daily as the virus took hold. Yet it was even harder, in that strange spring of 2020, to understand that beyond this crisis was another even more deadly truth: that the world was withering at the hands of humanity, that the virus of climate change had taken hold, and that, in spite of all appearances in those March and April days, the natural world was expiring around me. Rachel Carson begins her great and seminal book, *Silent Spring*, with what she calls a 'fable for tomorrow', a picture of a town where 'all life seemed to live in harmony with its surroundings', a place in the midst of a verdant land, where the abundance and variety of the birdlife was a thing of wonder. Then one day, she writes of 'a strange stillness':

> The birds, for example – where had they gone? Many people spoke of them, puzzled and disturbed. The feeding stations in the backyards were deserted. The few birds seen anywhere were moribund; they trembled violently and could not fly. It was a spring without voices.[7]

Turning those pages in the midst of all this awakening made Carson's words seem like science fiction. I thought of Richard Jefferies's book, *After London*, an

---

[5] S. Street. Unpublished diary entry.
[6] Ibid.
[7] R. Carson, *Silent Spring* (London: Penguin Books, 2000), pp. 21–2.

apocalyptic fiction from the nineteenth century in which humankind vanishes from urban streets, and the natural world takes over. Here was a new scenario, and this time, it was not fiction, but a vision of how someone in the not-too-distant future, sitting right where I was sitting, might experience this garden. Sound, like life itself, is fleeting. I've written many books about the aesthetics of sound, the poetic nature of audio; but it dawned on me there and then that any book I might enter upon that dealt with listening in the context of environment must demonstrate how the witness of changing sound patterns, locations and behaviour, can be part of a recording of understanding as we come to terms with what sound and the ability to experience it means as a symptom of existence. This IS about aesthetics, but more specifically it is about the aesthetics of existence and survival, not just of human life, but of Place, however, we choose to interpret that word. The cities were already silent, as Jefferies had prophesied. The spring trees opening around me held sounds that could never be taken for granted again.

Weeks after I recorded my garden songs, when some sort of 'normality' seemed to be reasserting itself and flights resumed from the airport, we noticed the roar of aircraft taking off in a way that drew attention to itself. We had been without them just long enough to become accustomed to a new, old sound world. But in those first days of the resumption, we noticed something else. As the first planes took off, the trees erupted with birds, flying in panic in circles before alighting again in a twitter of complaint. It dawned on us that many of these had been born into a world without jet flight, a world where the only sounds from the air were their own kind. Suddenly, something monstrous was bellowing as it rose into the sky above them. It was a sharp learning curve.

Before that happened, before the engines came back, we learned to listen in a new way, partly because the layers of audio grime had been removed, and partly because deep down, we knew this was temporary, that things would not last like this, so we came to value it, rather like an unrepeatable performance. Those birds I heard in my back garden were singing against a backdrop of other birds, of the rustle of spring leaves and the suburban soundscape of the neighbourhood: neighbours, children playing, a distant radio. Some said that the volume level of birdsong decreased at that time, because the birds did not have the roar of the world to compete with. There is always a sonic context, and in noting the focus of listening, we subconsciously register the scenery. In the late 1930s, when Ludwig Koch and E. M. Nicholson published their revolutionary *Songs of Wild Birds* and *More Songs of Wild Birds* as disc and book sets, their recordings seem to have largely isolated the bird from the landscape. We discuss these ground-breaking projects and their significance in the next chapter, but for now, it is worth noting that the songsters in the spotlight on those recordings perform largely solo, and it is hard in many of the recordings to get a real sense of place (although in Koch's recording of the chiff-chaff, the tiny song is sonic-bombed by a nearby pigeon!). For the most part, though, we are directed to the main subject, without distraction, and this of course was the intention; these are specimens, sound artefacts, and for anyone already familiar with the birds and their habitats, the imagination could supply the rest, as Julian Huxley wrote in his introduction to the first *Sound Book*: 'As the nightingale's voice escaped from its ebonite prison under the touch of the needle and the scientific magic of the sound-box, I felt myself transported to dusk in an April copse-wood.'[8] It is the power of associative memory.

Words hold the moment, but as Huxley wrote, the moment relived even as a slight version of itself through recorded sound, can trigger the imagination that in turn leads to its expression in words. With sound technology, we are in possession of a luxury not afforded to the likes of Gilbert White, Richard Jefferies and the Argentine-born nature writer, W. H. Hudson. They could not play back their world. Listening for them was a matter of noticing, experiencing and committing passing time to memory and to the page. They knew that while our attention may be caught by a spectacular birdsong in a tree close by, nothing happens in isolation. Even rudimentary attention can identify that when a garden male blackbird sings in the cherry tree, he is answering a faint song from across the park, marking his territory and noting that other voice's message coming back to him as he does so, while around him, the ambience of the season flows and swirls in sounds that are graded like colours. We have already used that catch-all phrase, 'a sense of Place', and it is as much a part of the experience as the key sound itself. It can help us even when the foreground sound is NOT natural, but urban and industrial, because the mind chooses what the ear tunes to, and over, under and beyond our immediate soundscape, there are other sounds. It is a question of focus, as the eye may see distant hills beyond the urban chimneys. In poetry and fiction, the mystery of landscape, perhaps in particular woodland, often becomes a narrative device, or equally, a metaphor as in Robert Frost's famous poem, 'Stopping by Woods on a Snowy Evening.'

---

[8] E. M. Nicholson and L. Koch, *Songs of Wild Birds* (London: H. F. & G. Witherby, 1937), p. xiv.

For example, in her novel, *Housekeeping*, Marilynne Robinson writes that 'the deep woods are as dark and stiff and as full of their own odours as the parlour of an old house. We would walk among those great legs, hearing the enthralled and incessant murmurings far above our heads, like children at a funeral.'[9] The woodland life is parallel to us, sometimes a dark metaphor reflecting human fear of the unknown, sometimes a place of refuge and of consolation. Either way, to walk into a wood is to enter another room acoustic, another kind of mood music, to which we need to adapt. In her book *Surfacings*, Kathleen Jamie describes going into a wood when world events – the everyday news – became too much, to 'consider what to do with the weight of it all.'[10] At first, lost in a maze of Scots pines, she stands and adjusts, 'hearing nothing, the non-sound of one leaf dropping to join its siblings on the ground.'[11] It is rather like focusing on a new language. As the mind homes in on detail, a sense in the sound takes shape and the hearing sharpens: 'there fall the tiny tin-tack calls of birds foraging in the treetops, the race of water in a burn.'[12] So tuned does she become that even the sounds beyond perception become psychically audible, and sight and audibility merge, as a moth flutters into view for a moment, and then passes out of human experience forever. High above her, a plane passes in a different world. Gradually the wood weaves its spell, and what began for Jamie as thought, turns into something akin to an imaginative trance:

> Green ferns in the groin of an oak. Green moss cloaking a stone. Voice of a crow. Voice of a chiding wren. A smirr of rain too soft to possess a voice. Voice of the shrew, the black slug. Voice of the forest … Did you hear something move out of the corner of your eye? The same moth come back? Or another leaf falling?[13]

This is what we are about here; the recording of sound and the preservation of the emotion contained in the moment of first hearing it, communicated through text. Of course, a sound file is a text, just as is the written word; reading and listening are kin. We receive the information, and our imagination, aided by memory and experience, creates the pictures, forming the emotions. Whether we are recording with a microphone, monitoring through headphones or actively listening, focusing with all our being on the world around us, we must re-tune

---

[9] M. Robinson, *Housekeeping* (London, Faber and Faber, 2015), p. 98.
[10] K. Jamie, 'Voice of the Wood' in *Surfacing* (London: Sort of Books, 2019), p. 243.
[11] Ibid.
[12] Ibid., p. 244.
[13] Ibid., p. 245.

our receiver – the brain – and train it as an athlete develops relevant muscles to serve their cause. Listening is a matter of choice and decision-making, but we have the capacity to make it more a part of us. The composer and musician Pauline Oliveros said, 'Listen to everything all the time and remind yourself when you are not listening.'[14] Such honing of aurality may require a daily work-out at first, a regular meditation, but gradually it will become a habit, then the norm, until we have learnt to hear the world as much as we see it. Oliveros also said, poignantly: 'I remind myself to listen so that I may be here now even though now has already gone.'[15] The memory becomes a sound bank so that 'primary listening gives us relationship to place and continues to unfold through one's lifetime – literally and figuratively'[16] When Thomas Hardy played back the sounds of Bockhampton Wood in *Under the Greenwood Tree*, quoted at the start of this book, he was re-running recollected forest murmurs from memory; it may have been listening at a remove, but because that same sound – or rather a descendant of it – remains available in some form, we know what he means. Or perhaps we have never heard anything like that; perhaps we have lived all our lives in an environment where there are no forests, no woodlands. Thomas Hardy is read all over the world after all; I once met a professor of English from a university in India at a Hardy conference in Dorset, England. He told me that when he went home, he always took a carefully preserved very small branch from a beech tree with him in his luggage, keeping the leaves intact as long as possible. He said he used it to give his students an idea of what a wood might sound like: 'Imagine hundreds of these rustling.' We start with the individual sound, and then move out into the multitude, and in his case, the inspiration was in the communicative power of the written word to evoke the sound of a place through suggestion.

There are times when only the *actual* sound will do, when the audio picture is worth a thousand words. In Chapter 1, we discussed the famous 1924 broadcast of the Cellist, Beatrice Harrison, playing a duet with nightingales in her garden, and the sensation it caused in a world still becoming used to the transmission of sound to a public audience. Over many years, this remarkable moment spawned a BBC tradition of outside broadcasts to hear the nightingales each May, as a

---

[14] P. Oliveros, 'Deep Listening: Bridge to Collaboration' in *Sounding the Margins: Collected Writings 1992–2009* (Kingston NY: Deep Listening Publications, 2010), p. 28.
[15] 'The Nature of Listening' ibid., p. 249.
[16] Ibid., p. 248.

harbinger of summer. Then, on 19 May 1942, during a live transmission, another sound became apparent. Some 147 RAF Lancaster and Wellington bombers were on their way to Germany on a bombing mission. The bird sang on, and behind it, the low drone of war planes created a dark accompaniment. The broadcast was ended for security reasons, but recording continued, and the resulting product still exists. There is no way such an event can be replicated in words; it is beyond the power of description to convey the poignancy and power of such a juxtaposition of sound. Sometimes the event overtakes the broadcasters' intention, and the result is a sort of sound poetry that is unique and irreducible. The great radio features producer, the late Piers Plowright, once told me that he believed there must be a god of location recording, a kindly angel that makes the church bell ring at just the right moment, or the bird sing in counterpoint, even when spiritual darkness descends. At the time of writing this, in early March 2022, the first signs of spring are showing in the English landscape. Soon, the trickle of birdsong already present will swell as the leaves and buds burst and the nesting season begins. Far from my awakening garden, Russian forces are invading Ukraine; there is a humanitarian and refugee crisis of vast proportions, and the idea of a world war seems closer than almost at any time in my life. Tanks are mustering outside Ukrainian cities, bombs and missiles are falling throughout the land. It is cold there, and snow is falling. But somewhere, in the midst of the death and suffering, is there birdsong? Some visitors to the site of the Second World War concentration camps speak of a strange silence. On the other hand, I think of Isaac Rosenberg's great soldier's poem of the First World War, 'Returning, We Hear the Larks':

> Sombre the night is.
> And though we have our lives, we know
> What sinister threat lurks there.
> Dragging these anguished limbs, we only know
> This poison-blasted track opens on our camp –
> On a little safe sleep.
> But hark! joy – joy – strange joy.
> Lo! heights of night ringing with unseen larks.
> Music showering on our upturned list'ning faces.
> Death could drop from the dark
> As easily as song –
> But song only dropped,
> Like a blind man's dreams on the sand

By dangerous tides,
Like a girl's dark hair for she dreams no ruin lies there,
Or her kisses where a serpent lies.[17]

Worthy of comment here is Rosenberg's choice of bird; we associate larks with morning song, and as Michael Guida mentions, 'for many officers and ranks the lark heralded the morning stand-to, the nightingale the evening one'.[18] Was the poet using licence here? Or did he mistake a nightingale for a lark? Possibly, although another answer might be, taking the description at face value, (especially that phrase 'heights of night'), that perhaps he was hearing a woodlark, which do sing at night, and from high up, and are more common on the continent than in England.[19]

To return to the work itself; like all great poetry, there are layers of meaning here: the sheer innocence of the birdsong, the non-partisan music that seems to sing as a consolation for the tired returning foot soldiers, while in truth, their song may also bring just as much joy to the enemy, who can also hear it. Beyond that, there is an innocence in the soldiers' response. For a moment, they are transported, before the reality of their situation returns. They are touched by an instant of grace, but it is dangerous to dwell on it too long, to allow the heart to soften too much, for that path leads to madness and death in their circumstances. Yet beyond this, there is that magical first reaction:

Lo! heights of night ringing with unseen larks.
Music showering on our upturned list'ning faces.[20]

There is the sound, conveyed through the words, of the high songs of birds 'showering' down on them as they look up. In those two lines, we 'hear' what they heard, and we have a sense of what it must have been like to hear it. A recording would have been eloquent, and would provide the evidence, while Rosenberg, in two extraordinary lines, lifts the aspiration of men to a state transcending their situation. 'Showering' is perfect, full as it is of sound, but also benediction, like gentle rain on a parched field. The mind imagines the sounds of shells and gunfire somewhere behind them, heard, but for a moment

---

[17] I. Rosenberg, *Collected Poems* (London: Chatto and Windus, 1974), p. 80.
[18] M. Guida, *Listening to British Nature: Wartime, Radio, and Modern Life, 1914–1945* (New York: Oxford University Press, 2022), p. 40.
[19] One of the passages of Olivier Messiaen's *Catalogue d'Oiseaux* is a kind of dialogue between woodlark and nightingale. I am grateful to Richard Mabey for his thoughts on this.
[20] I. Rosenberg, *Collected Poems*.

overridden by this new music, just as the fragile nightingale song defied the brutal throbbing drone of the bombers overhead. It is a selection of words that has the capacity to ignite sound, meaning and feeling simultaneously. In so doing, poetic writing captures the imagination through precision, suggestion, and with them, a silver thread to the mind of the reader. Rosenberg was dramatizing a moment that would have been familiar to many soldiers during the First World War; moreover, it was often the invisibility of the birds themselves that added to the poignant strangeness of the beauty, as Guida has explained:

> Because of the constrictions of the trench system and because of avian behaviour, birds were more often heard than seen. In fact, the two birds that received the most attention from soldiers were the most invisible. The sky lark would trill far above the fighting, lost in the sky, and the nightingale's shyness in the undergrowth, and tendency to give evening performances, guaranteed its illusiveness.[21]

Away from war, this experience of sound through words has often been what language has preserved at its most creative. It is the tool through which we convey our thoughts, and by refining it, we increase its capacity to work for us more eloquently, and this extends to the vocalizing of words, phrases and sentences. To make our oral argument more powerful in discussions, we pause, we emphasize, we change volume, pitch, tone, timbre without even consciously doing so. We strive for the 'right word' and the right word rings like a bell when it is struck. Poets using assonance, alliteration, rhyme and onomatopoeia, fire salvoes of sound to heighten emphasis and turn the invisible thought into audio and thence into pictures in the mind. Put concisely, in the words of Peter Levi, one way of looking at it is that 'poetry is language heightened by insistent sounds or repeated rhythms.'[22] One of the most famous passages in all English poetry, the 'Skating' sequence from William Wordsworth's autobiographical poem, *The Prelude,* contains the oft-quoted line:

> All shod with steel,
> We hissed along the polish'd ice[23]

---

[21] M. Guida, *Listening to British Nature: Wartime, Radio and Modern life, 1914–1945* (Oxford: Oxford University Press, 2022), p. 20.
[22] P. Levi, *The Noise Made by Poems* (London: Anvil Press, 1977), p. 30.
[23] W. Wordsworth, *The Prelude, or Growth of a Poet's Mind* (Oxford: Clarendon Press, 1959), p. 27.

Little wonder it is so well known, and used to demonstrate onomatopoeia in schools and in creative writing workshops. It is a string of little textual explosions strung along the lines on the page, detonating even as we read it silently, 'heard' and begging to be verbalized. The skates become audible, and the hard crispness of the moment is conveyed. Even the word 'along' is working with the other sounds to sharpen the effect. 'Across' might have sufficed, but it is softer, and while the 's' in it is longer, and more 'hissy', and so could perhaps have tempted the poet, the 'ong' sound rings like the echoes coming back from the dusk, a resonance that widens the attention beyond the individual to the place itself. William Wordsworth is amongst the finest examples of poets for whom sound was key; we shall revisit him later in this book, and explore some of his sonic techniques. Yet it is not only in poetry that we encounter this music; it happens in everyday speech, when overtaken by emotion and inspiration, grief or anger. Without knowing it, language can become transcendent in the service of passion, seeking a new precision and force. We may not craft our words so consciously, but meaning and feeling can make us all poets when circumstances require it.

The BBC's Beatrice Harrison/nightingale event in 1924 drove the technology to catch up, as content and circumstance so often do. A few years earlier, the idea of capturing such a moment would have been impossible, unthinkable. It would have come and gone, as sound events always did, and do. The only way to preserve, if not the sound itself, then at least the experience of hearing it, would have been through the written word. In the garden of Keats's house in Hampstead, north London, there have been many hopeful sound recordists, keen to catch a long-lost ambience, an evocation of the poet's time when he heard *his* nightingale and immortalized it in verse. As Angela Leighton has written:

> 'If actual sound is itself a transient passenger, invisible and always to be interpreted by the ear, how much more acute is the strange interpretability of sound in the written word, the ghost effects of which are built into its workings. For all reading is a matter of hearing things, in both the literal and the ghostly sense of that phrase.[24]

When we read Keats, we may 'hear' the birdsong imaginatively, while perhaps around us, the world plays its continuing soundscape, just as it did when the bombers accompanied the nightingale in 1942. While reading the signals, we

---

[24] A. Leighton, *Hearing Things: The Work of Sound in Literature* (Cambridge, MA: Belknap Press of Harvard University Press, 2018), p. 5.

have the capacity to subjugate the sounds that surround us, be they the train or the bus to work, aircraft flying overhead or domestic rummaging and voices. We can 'tune' our attention to the text, just as we are able to focus on one voice among many, say in a convivial and noisy party, a reception or waiting room. As Garrett Stewart has written: 'When we read to ourselves, our ears hear nothing. Where we read, however, we listen.'[25] Common to both direct sonic witnessing of an event, and the 'silent' transmission of sound through text, is often a pictorial element. While reading a book or a poem, the eye sends messages to the brain, which then turns these printed symbols into images. We are all recorders, and everything around us, when it comes down to it, is a text, whether or not we exercise our power of selection of one sound over another. Because a microphone has no mind, it does not select, and so we have an unfiltered recording of great power, where the whole sonic moment is preserved. The sound is what it is, but it is the individual who hears it, each single person, from their own physical, cultural and emotional perspective, interpreting it through the filter of national culture and identity. The lark, for example, has many meanings, and its song evokes different responses in various nationalities. For someone of Croatian nationality, for example, the sound holds a particular layer of interpretation as the national bird of their country. We each hear our own birdsong through our circumstances, location, background and being. It remains the same recording, but there can be multiple responses, only communicable through the secondary media of words or music.

Time and again we come back to making meanings, reading the sounds through the filter of personal perspective. To truly record an essence, we would need to have total and complete understanding, but this is where species divide and move into their own evolutionary private rooms. Where birdsong is concerned, we can learn from science about speed and frequency, habits and seasons, pitch and the formation of a syrinx, yet to understand how much on the edge of a sea of sound we really are, we should remind ourselves how much sound we actually miss through frequency limitation, sounds that hold deep and important significance for many other species, even those closest to us, like dogs. At the other end of the spectrum, low-frequency sound location can aid migrating birds by reading earth sounds as played through weather and geography: flat plains, mountain ridges, shorelines and marshlands, modulated by atmospheric changes. The

---

[25] G. Stewart, *Reading Voices: Literature and the Phonotext* (Berkeley: University of California Press, 1990), p. 11.

essences we convey through sound or word recordings seem, in the light of such thoughts, to be impossibly clumsy and lacking in the ability to interpret the nuances and subtleties of such communication. Among the many elements that separate us from the natural world is time, frequency and metabolism. In his book, *Being a Beast*, Charles Foster expressed it well: 'If, like many birds, you can hear sounds separated by less than two millionths of a second, you'll know the baroque complexity of apparently bland birdsong.'[26] We may have our seemingly perfect recording, congratulating ourselves on the capture of a sound on the very fringes of perception; we may equally have little difficulty in interpreting some of the more extreme emotional moments in the lives of birds, a scolding blackbird or angry attacking magpies, for example, but to really gain some idea of what is going on in the more complex avian sonic world, we need to slow the tape right down. The navigational signals going on in the murmuration of starlings, for example, or the quicksilver messaging of swifts, seemingly flying almost at the speed of sound itself, hint at something beyond us. As Foster writes: 'The acutely discriminating bird hears what I'd hear if I turned the speed of the birdsong right down. I can probably hear two sounds as distinct if they're around two hundredths of a second apart. The bird's getting in one second what it would take me about two and three quarter hours to hear.'[27]

Humankind has sought to emulate avian life since it was capable of expressing itself and probably before that. It is no consolation to know that birds cannot write books, because the next realization is that they do not have to. There is something else for which they have the code, that we seek to capture, be it in text or song. Sound is the key to the door, as well as the memory of what lies beyond it. It is also the layer upon layer of an imposed and increasing cacophony in the modern world that all but obscures the lost world, which the language of poetry and lyrical prose has sought to regain. We may use our technology to record the sound of the world, but in so doing, unless we at least seek an understanding, we remain outside observers. It is when the natural life around us enters our consciousness, as well as our microphones, that we come close to turning the key to the door into the secret garden. We are responders to where we are, to place and circumstance just as are the flora and fauna of which we are a part. What we hear of the world is the presence of things as they interact with other things. When we listen to the wind blowing, or the rain falling, we are not

---

[26] C. Foster, *Being a Beast* (London: Profile Books, 2016), pp. 189–90.
[27] Ibid.

simply hearing the sound of air moving, or precipitation, but those things in conversation with physical objects such as walls, gutters, shrubs and trees, the visible relative and apparent permanence of objects giving voice to unseen and transient elements. David Haskell wrote in *The Songs of Trees*:

> We hear the rain not through silent falling water but in the many translations delivered by objects that the rain encounters. Like any language, especially one with so much to pour out and so many waiting interpreters, the sky's linguistic foundations are expressed in an exuberance of form.[28]

We remain would-be translators of the natural world, as well as recorders of evidence. If some of the writers from history discussed in this book had had access to tape or digital recorders, we would have been the beneficiaries, in that there would have been auditory access to the distant past for us to share, analyse and discuss. But for themselves, I suspect that when it came to expressing the way those sounds made them feel, those same writers would still have set their impressions down in words.

---

[28] D. G. Haskell, *The Songs of Trees: Stories from Nature's Great Collectors* (New York: Viking, 2017), p. 4.

3

Ludwig Koch and the music of nature

The work of the German-born sound recordist, author and broadcasting personality Ludwig Koch (1881–1974), once a household name to the British public in the years immediately before and after the Second World War, is a bridge between modern technology and the descriptive literary heritage from which our idea of natural history sound originated. His principle field became that of radio, itself, particularly in its first half century, a strongly literary medium. Transcriptions of broadcast talks often appeared in print through journals such as *The Listener*, and many popular series spawned their own books, such as *In Town Tonight* and *The Naturalist*. Thus, Koch and other natural history broadcasters such as James Fisher, Peter Scott and Julian Huxley became well-known personalities to the listening public. It is partly for this reason that Koch earns a chapter in this study. The other major factor is that the literary nature of his broadcast work produced a form of natural history programming that combined location sound, often obtained with great technical difficulty, with a commentary by Koch himself that was highly evocative, idiosyncratic and at times poetic.

This chapter claims Koch as a link between the development of sound recording and the written word. There is yet another reason why this should be so, and it is the most significant of all; it is in Koch's creation with E. M. Nicholson, of a multimedia, mass produced tool which he called 'Sound-Books', of which there were initially two, the first produced in 1936 and the second a year later. It is worth taking some time to examine the form and content of the recordings, isolating as they do the bird in question like a solo singer on stage in a spotlight, in conjunction with a written text. No study of the mass impact of sonic bird literature can be without the acknowledgement of Ludwig Koch's place in the media world. His work looks both forward and backward within the continuity of avian auditory studies. The sound-books of Koch and Nicholson consisted of a box set with a number of shellac discs and an accompanying text, designed to place the recordings of the birds on the records into a context, and supply background information and illustrations. The books that accompanied the records were complete works in their own right, fully illustrated with photographs and some exquisite watercolours, whilst offering insights into the making of the discs, and the technology involved, which would have been revelatory at the time. These texts provide a rich combination of scientific observation and vivid description, as in the written section on the goldfinch:

> The gay, easy liquid song of the goldfinch, full of melodious phrases, is uttered as a rule in snatches of three to four at times up to ten seconds without a pause. The song is entirely free of harsh notes ... Fruit trees, poplars, walnuts, and to some extent oaks and yews ... from these, the goldfinch loves to sing.[1]

From the start, it is clear that the two 'sound-books' were intended to be part of a series. The phrase 'multimedia' was decades away, yet the packaging of these elaborate artefacts offered them to the public under the banner of a cross-over, selling them as 'Text – Sound – Pictures.' When the second sound book was published in 1937, the format was identical to the first, although with the addition of a third disc. The accompanying book built on the first by offering details of 'the new and fascinating technique of recording bird-song, with description of the birds recorded and how the songs can be recognised,' featuring examples of song 'from heath, park and woodland: the voices of the skylark, wood-lark, mistle-thrush, blackcap, curlew, and over a dozen other birds.'[2] The price was fifteen shillings, that is to say seventy-five pence in modern UK decimal currency; however, to put this into context in Britain at the time, average annual earnings were under £160.00.[3] It therefore becomes clear that products such as these were aimed at a highly select potential purchaser.

Had the war not intervened, this project would almost certainly have extended further, and indeed, Koch did develop other commercial sound publications, one with Julian Huxley, published by *Country Life* magazine in 1938, called *Animal Language*. As with the bird books, this was a combination of text, in the form of a 'coffee table' style book, with two discs bound into the back cover. These took the concept of wildlife recording beyond birds into the broader animal kingdom, among them dingos, seals, camels, crocodiles, hippos and many others. Thus, here was a product for the home bookshelf that offered

---

[1] E. M. Nicholson and L. Koch, *Songs of Wild Birds* (London: Wetherby, 1937), p. 83.
[2] Ibid., Dust jacket note.
[3] Measuring Worth, 2022. http://measuringworth.com/ukearncpi (accessed March 2022).

the general public access to sounds that they probably would not hear in the original habitat, with an explanation of meaning: evidence, emotional context and interpretation together, content that Koch would later provide so memorably in his broadcasts. And in the text itself, there is the blend of science, observation and the poetry that marked Ludwig Koch's own radio commentaries:

> Our little planet provides a speck of sound in a silent universe. The music of the spheres is a fable. In the cosmos at large there are formidable manifestations with which we instinctively associate violent sounds: explosions sending streamers of incandescent matter tens of thousands of miles above the sun's surface, comets and stars hurtling at enormous speeds through space, even collisions between suns greater than ours. But to speak of sound in such connections is meaningless. Sound as we know it demands, not only a disturbance of matter, but also more matter in a suitable form for the transmission of the disturbance, as well as the ear and brain of a living animal for its reception and its experiencing as sound. Neither in the depths of empty space nor in the absence of living beings of the same general nature as those inhabiting our earth have we the right to speak of sound.[4]

The records give us this sound, but the text is designed to capture the imagination. Access to the wild through recordings was a revolution, a door-opening, but what had not changed was the presence of the written word as a generator for context, meaning, excitement, awe and understanding.

Ludwig Koch was born on 13 November 1881, to an affluent and well-connected Jewish Frankfurt family. Music was from the start the centre of the family's interests, and the boy grew up within the social milieu of a rich German cultural society. From the start, young Ludwig had a naturalist's eye and ear for detail, and his recollection of the celebrities who came into his purview remained vivid into old age. In 1889, when he was eight years old, his father bought him and his brother the gift of two Edison phonographs and boxes of wax cylinders from the Leipzig Fair, and life changed forever.

Other than music, the family home was full of animals: reptiles, monkeys and birds, but not content with having immediate access to the sounds of nature, Ludwig would escape with his recorder to Frankfurt Zoo, just fifteen minutes from his home, whenever possible. Koch was growing up at precisely the time when the technology of sound recording was developing at an astonishing rate;

---

[4] Ibid., p. 7.

Emil Berliner created what was to become known as the gramophone, and in 1906, a family friend, Max Strauss, founder of the Parlophone company[5], presented Ludwig with a disc recording machine, 'with which I made a good number of open-air recordings in the field, garden and woodland. The results were not too bad, but there was no way of preserving those wax recordings, since the few existing pressing factories were not interested in this kind of work.'[6]

For a time, music led the career path. Koch was a fine singer, of professional standard, and as a young man he spent many months in Bayreuth, moving into the circle of Siegfried and Cosima Wagner, but with the coming of the First World War, he was assigned to German military intelligence. In 1918, he became the chief delegate for the repatriation of the French-occupied zone of Germany and worked for the government until 1925, when he was commissioned by the German subsidiary of Electrical and Musical Industries (EMIs) to develop a cultural branch of the growing gramophone industry, enabling him to revisit his fascination with wildlife recording and gaining access to the newest technology, which he began utilizing from 1929. The years 1928–1936 were significant for Koch and his ideas, notably the development of the multimedia concept of sound-books:

> In the eight years of work with EMI, I was able to introduce completely fresh developments in the use of the gramophone for educational purposes. These included the publications of two series of records and accompanying texts entitled 'Two Thousand Years of Music' and 'Music of the Orient.' I also published a monthly magazine, which contained contributions by Hindemith, Stravinsky, Madame Montessori, and others. As far back as the late 'twenties I was already thinking of the value of a book on birds and animals, illustrated by good recordings instead of those musical notations and curves which meant nothing either to a scientist or to a bird-lover. The translations, too, into such words as *tu, tu, tu* or *tse, tse, tse* will never bring to the ears of the average listener the sweetness of the song of the wood-lark or the characteristic note of the marsh-tit. Thus in my mind was born the sound-book, a combination of text, picture, and sound, the last supplied by means of gramophone records attached to the book. The success of this idea was so great in Germany that within four years I was able to publish eleven sound-books, three of them on animals.[7]

---

[5] Max Strauss founded Parlophone (originally 'Parlophon') with Heinrich Zuntz of Salon Kinematograph, in 1902. The first Parlophon [sic] commercial records were pressed in 1910. Strauss was one of the many well-known and powerful members of the Koch family's Frankfurt circle.

[6] L. Koch, *Memoirs of a Birdman* (London: Phoenix House, 1955), p. 16.

[7] Ibid., p. 25.

He was commissioned by the cities of Cologne and Leipzig to make sound-books based on their various ambiences, 'for nearly every city has its characteristic sounds'. It was an idea that he later developed in Paris; his fame as a sound recordist reached high places, and his recordings were played on numerous German radio stations as the medium gained ground – and listeners. Yet life for a Jew, however well-connected, in Germany during the 1930s, was bound to become difficult, and Koch was an outspoken critic of Hitler and Goebbels, to the extent that an overheard conversation in late 1935 led him to be questioned by the Gestapo. As it happened, he had already accepted an invitation to give a series of lectures in Switzerland, beginning in January 1936, at the end of which he contacted his employers at the Gramophone Company in Germany to discuss arrangements for his return. He was advised to remain in Switzerland, and at a conference, he met the director of the company, who pointedly suggested that he did not come back to Germany. In fact while he had been away, a warrant had been issued for his arrest. It was a life-or-death decision, and, as it turned out, a key turning point in his career. Among those in Lausanne for the conference was Louis Sterling, Managing Director of EMI, who invited him to England; thus, on 17 February 1936, he landed at Dover and began a new life, bringing with him – if not many of his prized recordings – at least numerous ideas and concepts for the furthering of environmental sound.

Koch's reputation preceded him, and there was considerable enthusiasm for an English version of one of his sound-books, but in his hasty migration from Germany via Switzerland, he did not have any examples to demonstrate this unique hybrid. As luck would have it, one of the managers at EMI possessed a copy of his last German sound-book, *Gefiederte Meistersänger* ('Feathered Mastersingers'). With this as a sample, he was introduced to Julian Huxley, at that time the newly appointed secretary of the Zoological Society of London, a founding member of the World Wildlife Fund and the first director of UNESCO. Huxley provided the addresses of a number of book publishers appropriate to the subject matter, and Koch tried to sell the idea around London literary houses, initially to no avail; probably the mixed media concept, and the costs involved in creating a box set, as opposed to a single book that would sit snugly on a shelf, was a discouragement. Whatever the reason, first responses were overwhelmingly disappointing, but in the end, it was a letter from his last collaborator in Germany, the ornithologist Oscar Heinroth, that produced a result, in the form of H. F. and G. Witherby, a publisher specializing in natural history. It was an inspired contact; Harry Witherby was himself a keen and knowledgeable ornithologist, and a visit to the firm's London office produced the first positive reaction to Koch's ideas. It is interesting that this blending of text and sound, so enthusiastically and widely adopted in his native country, should have had such a hesitant birth in Britain, but with the Witherbys on board, the first English sound-book began to take shape. Harry Witherby introduced Koch to the writer Max Nicholson, who had recently been commissioned by the firm to write a book on British birds, with the working title, *Songs of Wild Birds*, and it was agreed that this should incorporate two 10-inch discs of Koch's recordings. By April 1936, Koch, with an HMV-recording unit and a number of engineering staff, had begun working on their first recording, a song-thrush, made at Green Street Green, near Bromley in Kent. Ludwig Koch's autobiography gives a vivid and detailed description of subsequent location recordings as they were made over the coming months. The speed of production of this product, for which there had been no manufacturing precedent, was remarkable, and by October, Koch was giving a launch talk at the Royal Society of Arts in London, broadcast by the BBC. The response was immediate and dramatic. Members of the Royal Household wrote to Koch, and he was invited to Downing Street by Neville Chamberlain. Chamberlain was a very keen naturalist and ornithologist, but the situation in Europe was deteriorating, and the timing of the visit led to some controversy, as Koch himself explained:

> Mr and Mrs Chamberlain took a great interest in my work, and after the Munich meeting a prominent Swiss paper attacked me for taking up the Prime Minister's time with listening to bird-song instead of to the voice of the threatened people of, Czechoslovakia. The paper even claimed that Mr Chamberlain had left an important meeting in 1939 to listen to the unusual song of a blackbird in the garden of 10 Downing Street, who mimicked the repeated notes of a song-thrush. I do not believe, however, that either his interest in my work or this incident had any influence on the development of history.[8]

The public response, despite the cost, was equally encouraging, and Koch soon realized that the cross-over potential of this interdisciplinary project in retail terms increased its desirability. 'It was encouraging for me to see the sound-book in the windows of book shops and gramophone shops, and it received tremendous publicity throughout the Commonwealth and Empire.'[9] In addition

---

[8] Ibid., p. 41.
[9] Ibid.

to the first sound-book of birdsong, work was ongoing during 1937 on a second, as well as a new project proposed by George Witherby of Koch's publishing firm, who was a keen huntsman. Witherby proposed a project to be called *Hunting By Ear*, which described in text and on disc, the sonic process of fox hunting. The text on this occasion was supplied by Michael Berry and D. W. E. Brock with Koch supplying the recordings from a hunt at Everleigh near Marlborough in Wiltshire. The fact that a sound-book in the same format as *Songs of Wild Birds* could be considered a commercial proposition tells us much about the times, and the audience context in which such a work would be viewed as viable and appropriate. In fact, evidence of its success may be understood when it is noted that after the war, it was reprinted as late as 1949, and even in the twenty-first century it was transferred to paperback and CD, receiving some enthusiastic responses. While the subject matter and attitudes behind it may be less than attractive to many of later generations, there is no doubt that, viewed as a document and a part of a rural archive of England, Koch's recordings capture unique sound pictures. Disc one sets the scene: 'The Horn', 'The Huntsman's Voice' and 'The Whipper-In' (The Whipper-In is the assistant to the huntsman who brings straying hounds back into the pack). The concept of text and record interlocking in meaning and use was encouraged with explanatory notes, placed inside the smart red cardboard box within which book and discs were housed:

> Before playing the records, look at the book to get the plan of HUNTING BY EAR. Look at the notes on page 91 and then play through Record 101A Then read through the notes on pages 100–102 and play Record 101B. Next, while playing Records 102A & B, have the book open in front of you, page 103 onwards, and let your eye skim the italics.[10]

This part of the text was intended as an aid to understanding the context, setting the scene in the mind's eye, for example, '0–10 seconds: *The huntsman rides up to the covert-side with his hounds. His two whippers-in have gone on for a view.*' There then follows non-italic text explaining meaning particularly for any listeners unfamiliar with hunting, its practices and traditions.

> Before playing them a second time read all the commentary; you will then know what you have missed, and whether, when hunting by ear, you have guessed rightly what was happening.[11]

[10] M. Berry, D. W. E. Brock and L. Koch, *Hunting by Ear* (London: H. F. & G. Witherby, 1937), insert slip.
[11] Ibid.

Thus, the whole production may be seen as a training tool in the art of listening actively, of focusing and targeting sounds through concentration. The book, had a narrative section written by Berry and a commentary to the records supplied by Brock. The text was integral and inseparable from the recordings: 'Without the text and visual illustrations much of the value of the records will have gone, and it is to the text first and *then* to the records that the reader must turn before he can appreciate the full scope and purpose of this sound-book.'[12] Throughout, Berry focuses on the music of not only the hunting horn, but the huntsman's cries and the sounds of the hounds themselves, what he calls 'hound music'. These provide an audio text in itself that can be read and interpreted as a barometer of a successful hunt. It is, he writes, 'the elemental sound – the salt that alone can give the savour, the mercury that alone can supply the weight. The successful fox-hunter, then, is he who listens to the cry, who can detect changes in its tone and volume, and who can keep altering his own course on horseback so that he never loses touch with the body of the pack.'[13]

For Koch, the project presented a unique challenge; here, instead of a fixed point upon which to aim his microphone, there was a moving sound-target, crossing the countryside at high speed. At first, he was doubtful, especially after being taken by a member of the Witherby family to witness a hunt: 'I was totally ignorant of a fox hunt ... I could not imagine how it would be possible to record a hunt without special staging – which the publishers definitely did not want.'[14] The answer was to identify as far as possible the strategic points where sound could be more or less guaranteed, and the Everleigh hunt in August 1937 offered a promising location, given that it was in Koch's words, 'within rather confined territory.'[15] The result was not a continuous soundscape, but a series of sonic pictures, with which Koch and his publishers were very satisfied. 'We succeeded in making a series of first-class recordings of the horn, the huntsmen's voices, the whippers-in, and finally of the hunt. The fox and the hounds close upon it passed our microphones several times, even when they went through a pond.' *Hunting by Ear* was clearly aimed at a niche market, but it was an audience the publishers felt sure of, indicative perhaps of the business model that supported the success of Witherby's decision to publish sound-books overall.

---

[12] Ibid., dust wrapper note.
[13] Ibid., p. 25.
[14] L. Koch, *Memoirs of a Birdman*, p. 52.
[15] Ibid.

Measured against radio listening of the time, these records offered a number of unique advantages, not least in terms of relative sound quality, but also in the facility for the reader/listener to turn back the page and play sounds as often as required or desired. Today, we would only need to digitise the product in order to place it within a media world with which a modern consumer would feel totally at home, and for whom the whole principle would be at once familiar.

Thus, with the publication of *More Songs of Wild Birds* and the hunting project, two new sound-books were in the shops by Christmas 1937, with the Country Life publication of *Animal Language*, featuring recordings by Koch at London Zoo in Regent's Park, well on the way. The text for *More Songs* largely followed the model of the first book, with commentary at the end feeling at times rather like programme notes at a concert, to be read while listening to the 'music' itself. For example, against a time signature to enable the listener to place text and sound together, we have this explanation of the skylark:

> 0.0 As the recording starts, a skylark is heard rising strongly, pouring out the typical exuberant song. At times the volume of song drops a little, but after a little while it recovers full strength, such fluctuations being quite normal as the bird sings on the wing.
>
> 0.30 The singer is now getting quite high above the microphone, and the song is perceptibly fainter. It goes on retreating into the distance and by 0.45 is rather faint. During the next ten seconds the peak of the ascent is passed
>
> and
>
> by 1.00 the increasing loudness of the song announces that the lark is on the way down. He continues to get louder and finally breaks off quite close above the microphone at 1.18.[16]

The birds represented in the recordings were wide-ranging: blackcap, garden-warbler, wood-wren featured on one record, and crows, rooks, magpies and jays were grouped on the next, while the haunting sound of the nightjar's 'churring' sound ends another. Yet it was Koch's recording of the curlew that was most striking, and became his most famous 'capture', largely due to its adoption by Desmond Hawkins, producer of the BBC radio programme, *The Naturalist*, as the regular theme, used in place of a signature tune, at the time

---

[16] E. M. Nicholson and L. Koch, *More Songs of Wild Birds* (London: H.F. & G. Witherby Ltd, 1937), p. 82.

a normal opening to any radio programme. Yet as varied and successful as the second bird sound-book was, there was still an elusive sound that obsessed Koch, and which he pursued after most of the scheduled sound-hunting for the discs had ended:

> I was haunted by the thought of the tawny owl, and in the late afternoon I managed to locate the hooting spot of this night bird. We got into position, but whenever the microphone was placed, after a lot of strenuous work, the bird would fly five hundred yards further. We followed him for many miles, and presently I had the impression that the bird had settled down in a very dense wood. The car and the recording van followed as quickly as possible, none of us noticing in the frenzy of the chase that we had entered private ground. Then suddenly lights went on, and I could just hear a frightened voice telephoning for the police. We did not stay to await further developments.[17]

Possibly in the crew's mind was the fact that, in the climate of the times, an attempted explanation from a man with a microphone and a strong German accent, might not have helped the situation. It was to be another ten years before Koch was to obtain his much sought-for recording of the tawny owl, which he ultimately collected in Dorset. The example remains a potent one of the difficulties attending the technology of recording at the time. How much easier for a writer to listen and interpret solely on the page.

Koch's next project took him to new fields once more, with an invitation to collaborate with the Belgian royal family, and particularly the Queen, who was keen to learn the principles of wildlife recording. Koch spent many months working, instructing and recording on the Belgian royal estate near Brussels. A sound-book, *Oiseaux Chanteurs de Laeken* was close to completion, with a French text written by Koch, with publication scheduled for 1939. In the end, circumstances dictated that the project was delayed until the end of the war, when, in 1952, it was eventually produced and distributed in an edition of 20,000 copies, to Belgian schools.

We have already touched on Koch's desire and ability to isolate the bird he sought to record. To him, the sound was an artefact, a specimen; but there was also an element of giving the solo performer its due space. He recorded and listened to the world as music, whether it was a cityscape, a street performer (as in a recording of a Parisian busker playing the spoons, later parodied by Peter

---

[17] L. Koch, *Memoirs of a Birdman* (London: Phoenix House Ltd., 1954), p. 48.

Sellers for one of his albums[18]) or a rare bird. He went to enormous lengths to obtain the right focus, and he was clearly demanding in his requirements of those who worked with him. Considering the bulk and weight of the equipment at his disposal, including the tape machines, disc-cutters and miles of cable, all transported in the heavy Parlophone truck with reinforced suspension, as well as the sheer physical effort of negotiating terrains and setting up microphones, the wonder is that these recordings were achieved at all. Yet the tenacity and patience of the man led him to tireless lengths in pursuit of his goal. He was even known to instruct his team to 'flush out' competing birds from a neighbourhood, in order that the main performer should have the sound stage to themselves.

An unexpected benefit came in the latter part of the war with the development of more portable technologies, designed at the recommendation of front-line reporters such as Frank Gillard, to enable BBC microphones to follow the unfolding progress of allied troops across Europe and in the air. The 'Midget' recorder was indeed a revolution, and, while still bulky by today's standards, it freed journalists to provide many unforgettable eye witness accounts on location where the old 'mobile studios' would have had no hope of reaching. Through these years Koch continued to record, and to work within the BBC: 'During the war years it was happily possible for me to continue, using a portable recording outfit developed for training purposes, which proved very suitable for my special work.'[19] Nonetheless, he was not immune to the tensions of the time, and during 1940 he briefly suffered a period of internment as an enemy alien, but was released when, at the suggestion of Julian Huxley, the BBC's European Service and subsequently in the Home Service, agreed to employ him.

He began creating illustrated talks around his wildlife recordings, and his presentation became legendary, with its singing quality and its poetic style. Koch had a strong narrative gift, and his ability to paint pictures in words as well as recorded sound, made him a natural radio personality, sharing the mood and feeling of the experience through vivid scene setting and storytelling. This ability is perceivable in his written texts; anyone who has ever heard a Ludwig Koch commentary, cannot help but 'hear' his voice as they read accounts of his travels and 'adventures in sound recording' as he called them, for example, in

---

[18] P. Sellers, *The Best of Sellers* (London: EMI Records), MRS5157, Side 2, Track 4: 'Suddenly It's Folk Song', 1958.

[19] L. Koch, 'Adventures in Sound Recording' in *The BBC Year Book 1949* (London: BBC, 1949), p. 21.

this recounting of a trip on Horsey Mere in Norfolk, in search of the 'boom' of the bittern during war time, published in the *BBC Year Book* for 1949:

> One fine night our floating recording studio was in search of a hiding place in the swampy reed beds where I had often heard the bittern. On this night it did not boom a single time within recording range. After dawn, the wind rose to a gale, and BBC recording engineer Eric Hough and I thought the boat no longer safe for my precious recording gear. We moved to a more sheltered spot, but for two nights heavy wind and the rustling of the reeds made any attempt impossible. Then came a night when the wind dropped and the bittern boomed within recording range. 'Cut', I said quickly (meaning 'Start recording'), but in the same fraction of a second a bomber squadron began circling overhead. It was tantalizing in the extreme to hear the bittern without being able to record.[20]

Any field recordist who has ever attempted to capture natural sounds on location will have sympathy with Koch here, particularly with the benefit of the hindsight of all the subsequent years of increasing noise pollution and diminution of wild life habitats. Beyond that, however, there is the narrative voice speaking through the text, evoking the scene, and creating a mood that makes this mini-drama alive and real. It was a major quality in Koch's later work, and makes him a key part of the blending of sound and text in location storytelling. Having already created the multimedia form of the sound-book, it would only have been a small step for Koch to have added his own narrative to the discs, thus effectively developing the audio book. Perhaps, given the times and circumstances, Britain in the 1930s and early 1940s offered neither the time nor the place for such an experiment.

I have written elsewhere of Koch's BBC career.[21] Suffice to say here that when it came to his broadcast commentaries, there is little doubt that he used a cultivated persona, (again, a factor that made him the subject of affectionate parodies in the world of comedy). He was himself a performer, and presented to some extent a caricature of the eccentric German 'boffin' which endeared him to legions of radio listeners, although it was less enthusiastically welcomed by some in the more serious-minded scientific community. He was a sound pioneer, but also a populist with an eye on the commercial market, and the combination of these elements created a media personality who left a legacy that continues to

---

[20] Ibid., pp. 21–2.
[21] S. Street, *The Poetry of Radio* (Abingdon: Routledge, 2013), pp. 100–7; *The Memory of Sound* (New York: Routledge, 2015), pp. 123–4.

this day, while there is little doubt that his own awareness of his pioneering work, together with his ability to play to popular appeal, had its critics. For example, his BBC successor as the director of wildlife sound recording, Eric Simms, gave Koch his due in these somewhat qualified tones, in his book, *Wildlife Sounds and their Recording*:

> The indefatigable Ludwig Koch, without a background of ornithological training but advised by many naturalists, became 'a sleepless obsessive' in his pursuit of nature's sounds and was the first to describe the difficulties of wildlife recording and how in the end 'endurance was rewarded.' This unconventional and somewhat egocentric man pioneered the mechanical recording of the voices of birds and remained loyal to his discs even after new techniques offered greater frequency ranges and ease of operation.[22]

Koch's work, as we have discussed, anticipated the idea of multimedia as a tool for education, and his capturing of city sounds, rural pursuits and general atmospheres pre-figured environmental acoustic studies and field recordings, works designed to be listened to with full attention, as one would listen to a symphonic work. He was fascinated by the particular, but he was also strongly conscious of its context within the broader sound world. His recordings are both on occasion wide-screen and close up, as and when the circumstances required. His concepts, from the earliest sound-books in Germany through to his broadcast commentaries, featuring the sounds gathered on his travels and dramatized by his somewhat eccentric presentation, pre-figured much that has in the twenty-first century become the fabric of 'new' audio: field soundscapes, natural history actuality, the development of location recording, and new modes of capture and preservation of sound. The interactive nature of the sound-books, and his self-contained radio features involving elements of what we would now understand as podcast structures, show him to be in many ways the father of us all who seek to hold the fleeting moment of experience of sound through modern and evolving technology and play-back conventions.

His dream, from the early years, was that there should be a permanent national library resource for sound, and that audio should be treated in the same way as paper and other artefacts. Julian Huxley, with whom Koch collaborated on *The Language of Birds* sound-book, wrote in 1938:

---

[22] E. Simms, *Wildlife Sounds and Their Recording* (London: Paul Elek, 1979), p. 6.

> If he could have his way he would establish a great Sound Institute to deal with all aspects of the problem of recording interesting sounds. Dialects, folk-songs, thespeech and music of primitive peoples, the songs of birds, the sounds of other animals, the voices of famous men and women, the characteristic sounds of different industries and cities – all these and much else would be stored for the use of our own and future generations.[23]

It was an idea that obsessed Koch for many years, but in the financial climate of the 1930s and 1940s, there was little possibility of its realisation. Even his friendship with sympathetic royalty and politicians was not sufficient to establish a business plan as a foundation, as he wrote in 1955:

> I always had in my mind my old idea of a Sound Institute. I was disappointed when Neville Chamberlain explained to me that in this country there are no public funds available for undertakings of that sort, and that I should have to seek financial help from private sources ... I did not know of anyone who would support my idea of a sound institute for natural history, music, folklore, dialects, famous voices of the past, and so on, and to this very day I have not managed to advance one step towards the realisation of this work.[24]

It was possibly this sense of frustration that led Koch to donate his collection to the BBC. His bitter experiences in pre-war Germany had shown him that preservation and conservation were above all the first priorities. As it happened, he was not alone in his ideas. During the 1930s, a teenage boy in London, by the name of Patrick Saul (1913–1999)[25], seeking an out-of-print recording of a favourite piece of music, went to the British Museum, and was amazed when he was told that the Museum did not possess a collection of records. Ultimately this discovery led Saul to become instrumental in the creation of what was initially known as the British Institute of Recorded Sound (BIRS), later to be absorbed into the British Library as The National Sound Archive, finding its home on Euston Road, alongside the greatest documents of history and literature. Today, under the banner, *British Library – Sounds*, it houses an ever-growing public archive of over 90,000 recordings, encompassing accents and dialects, arts, literature and performance, classical music, popular music, oral history, radio

---

[23] J. Huxley and L. Koch, *Animal Language*. p. 1.
[24] L. Koch, *Memoirs of a Birdman*, p. 41.
[25] Saul became a mature psychology student at London University with a large personal collection of recordings. He was appointed the BIRS's first secretary at its foundation in 1948, and retired as its director in 1978.

and sound recording history, world and traditional music and sound maps; in short, it is the home of sound envisaged by Ludwig Koch from the start. So much of Ludwig Koch's work prophesied today's continuing interest in location sound, and much of the development of bioacoustics owes its genesis to his tireless and obsessive efforts in all weathers, often alone. This, combined with his magnetic radio personality, opened up auditory access to the natural world, and there are few in any of today's sound recordists who do not know his name, and acknowledge a debt to him. 1955 saw the publication by the Waverley Book Company of the huge single-volume *Encyclopaedia of British Birds*, edited by Koch, and with an impressive team of contributors, including James Fisher, Desmond Hawkins, Eric Hosking, Julian Huxley, Peter Scott, Eric Simms and Brian Vesey-Fitzgerald. In his introduction, Koch drew attention to the book's ground-breaking contribution to the field:

> Nobody has ever before attempted to produce an encyclopaedia such as the present one, which has condensed into one volume, arranged in alphabetical order, all the subjects on which those interested in bird life – from the amateur bird watcher to the serious student of ornithology – are likely to require information. In fact it can claim to be *the first of its kind*. [Koch's italics.] Moreover, it is the first book which, in describing birds and their characteristics, gives full consideration to their vocal performances. This subject is neglected in some bird books, while in others it is represented by musical notes or symbolic notations, which I have always considered not only inadequate but entirely misleading. In the present book I have been able to offer information drawn from my own life's work.[26]

It is indeed a sumptuous and comprehensive volume, with its almost six hundred pages studded with some forty full-colour photographic images, and, as Koch promised, vivid textual 'recordings' of its subjects' songs, conveying not only the fact of them, but something of the spirit and poetry, as in this entry on the goldfinch:

> The song of the goldfinch, a gay and liquid performance full of musical notes, is usually uttered from a tree or a hedge some 10 feet from the ground. The song is composed of three or four phrases delivered without a pause. The goldfinch is perhaps not so strictly territorial as some of our small birds, for it does not seem to mind the intrusion of another one in the vicinity and, indeed, they perform

---

[26] L. Koch (ed.), *The Encyclopaedia of British Birds* (London: Waverley Book Company, 1955), p. v.

wonderfully together in the spring. The song is heard at its best from March to June.[27]

The last recording made by Ludwig Koch for the BBC was in Somerset, in 1961, when, in his eightieth year, he captured in sound, a nest of young swallows. A year earlier, his services to wildlife recording had been recognized by the award of an MBE. In 1960, he created one more sound-book, for the most part using his BBC recordings. It was produced by the prophetically named 'Talking Book Company' in association with the publishers Methuen, once again anticipating trends still more than half a century away. A seven-inch, vinyl thirty-three rpm disc, was enclosed in a gate-fold sleeve with watercolour images by artist Richard Taylor of the birds included on the record, alongside explanatory notes and an introduction by Koch, which ended with a firm injunction to 'start birdwatching without delay.' Side one comprised recordings of the house, tree and hedge sparrow, the carrion crow, rook and raven. The second side featured the starling, blackbird, song-thrush and mistle-thrush. Thus, the record took advantage of the relative quality of vinyl, the portability of the small disc format, and the long-play format, enabling the inclusion of a range of birds that would have needed some four or five old-fashioned 78 rpm discs in the first sound-books. The recordings also included something else the first discs had not been able to boast, namely the voice of Koch himself, although anyone buying it in anticipation of hearing his colourful and poetic commentaries so familiar through radio, would have been disappointed; here, Koch limited himself to simple, monosyllabic identity statements and no more. One other benefit of the 1960 sound-book was the price: at eight shillings and six pence, it represented in relative terms, a far better deal than the first box sets of the 1930s.

There were to be other recorded testaments to Koch's work. In 1970, four years before his death, the BBC issued a 12-inch LP record entitled *Ludwig Koch: Recollections and Recordings*.[28] It was narrated by one of his main BBC champions, Desmond Hawkins. Side one was entitled 'The Man', and included accounts of many of Koch's recording expeditions, including an extended version of a radio programme produced by John Burton, the first broadcast in the Radio 4 series, *Listen*, on 14 November 1969. Side two featured some of the most famous Koch recordings from Surrey, the Norfolk Broads, Dorset, the

---

[27] Ibid., p. 255.
[28] This was in fact an extended sequel to an earlier BBC disc called *A Salute to Ludwig Koch*, which had been issued a few years earlier.

Scottish Highlands, the Channel Islands and Skomer, Shetland and Dartmoor. It was an immediate success, and *The Gramophone* magazine, in its December 1969 annual round-up voted it the Documentary Record of the Year.

It is important that this work is acknowledged here, pivotal as it is, as a meeting point between not only the blending of the abilities to record and to envisage, but in the realization that sound – and therefore radio, particularly in its formative years – is at its heart a literary imaginative text. No child who heard it will forget Koch's illustrated radio talk on recording Atlantic grey seals on the island of Skomer, off the Pembrokeshire coast, part of which was quoted from earlier. It is a text he reproduced almost verbatim in his autobiography:

> Suddenly a ghostly voice came to my ears, I did not know whence, but automatically I started recording. The longer I listened, the more certain I became that the age-old stories of mermaids had come true. The song I heard must have been brought to me by the wind from some rock where the seals were singing, but to me, listening enthralled in the darkness, it seemed like a corroboration of all the legends.[29]

Something in Ludwig Koch went beyond the technology of recording sound; he was a storyteller who understood that the printed word, be it in book form, or verbalized as part of a broadcast script, could illuminate the sound, and contextualize it, but at its best, also enhance its mood and drama. Sound and text share some key elements; notably, both provide information while allowing the imagination to create pictures. Ludwig Koch's work looked both back to the literary tradition of emotional expression evoked by listening, and forward to the multimedia world in which text and recorded sound interrelate.

---

[29] L. Koch, *Memoirs of a Birdman*, p. 99.

4

Murmurings: Call and response between bird and human

During the 1980s, I began a series of radio and literary collaborations with Desmond Hawkins, who had by then retired from the BBC, while still active as a writer and adapter of work by writers such as Thomas Hardy for radio. Hawkins was a sound man through and through, and as we have seen, a key advocate for the work of Ludwig Koch within the BBC. The development of portable recording overseen by him and his colleague Frank Gillard during the war years was crucial to opening up the first-hand experience of the natural world, and his writing conveyed the important contribution made by radio, utilizing this revolution in location sound:

> Many listeners have commented on their acquisition by radio of a form of knowledge that books cannot possibly give. The hatching of a greenshank, the singing of seals, the howl of a vixen, the booming of a bittern, the family conversation of swans at night, the angry piping of queen bees, and the gentle murmurs of eider duck – these are some of the fascinating sounds that listeners have heard in the last four or five years. And with the sounds, expert commentaries by distinguished naturalists.[1]

As a literary man, Hawkins understood perfectly that sound will always be linked inseparably to the imagination and memory, as creative writing has the capacity to suggest sound through language, opening in turn mentally formed visual images. He also knew how the writers he loved – Hardy, Jefferies, Hudson and Thoreau – used their pens as microphones, while his own generation of broadcasters were discovering how to 'write' onto disc and tape as they witnessed and sought to explain the natural world in which they had embedded themselves. We often talked about the early days, when radio was a new and democratising medium. Knowledge, music, thinking: all shared via sound, 'broadcast' like

---

[1] D. Hawkins, 'Nature Broadcasting' in *BBC Year Book 1951* (London: BBC, 1951), p. 33.

seed, to be absorbed into a wider public base of understanding. To draw a direct comparison between a microphone and a pen opens a complex discussion, because the mic., even in the hands of the most sensitive and perceptive recordist, can only capture what already exists, whereas the pen as we have seen, translates that presence into a contextualized thought or emotion. There is however, a form of translation, or rather data transference, through technology, from the original sound, by whatever means, arriving at the stage of playback. There is also an essence, about which we spoke earlier. The first transmissions and recordings were suggestions of a reality they were not yet fully equipped to mirror.

In the early years, this involved the tortuous process of printing onto hot wax, towards a more or less permanent and fixed product. Even with the so-called purity of digitally recorded sound, there remains the nature of the listener experience. Even latterly, a shrunken mp3 sound file would seldom do justice to what the recordist heard in the woodland, on the beach, beneath the waves or on the hilltop. A counter-argument says that the broadcast or recorded sound itself offers a sense of direct witness, be it a place, animal or human voice, whereas 'The written text … represents sounds that will have to be translated by the mind of the reader into actual or virtual sound. An image described in a text is sketched with a degree of abstraction that asks the reader to fill in the spaces.'[2] (This, some would say, may actually be seen as an advantage rather than a shortcoming, liberating the imagination, a merit in fact common to both radio and the printed word.) We create versions, interpretations: mnemonics are suggestible to the mind in capturing the spirit behind the murmurings. Across the centuries, the oral had become written, and now the printed or conversational word could be shared through this new invisible medium. In this chapter, I want to explore how, prior to electronic media, writers through time from various social strata have sought to share the sounds of the natural world, often reflecting the human condition in the process. As with broadcasting, developing in its various forms, there were two strands, which we might call 'artistic' and 'journalistic'. The former, we may identify as the product of writers by profession for whom the style and content forged a reputation within a canon. The latter largely originated from amateur writers, often country men and women, for whom an immediacy of experience 'in the field' communicated itself through their passion and experience, revealing a state of living while preserving a record of local

[2] Ibid.

customs and history. Also contained within it was a written version of 'location recording' that could flash with the truth and unfiltered honesty of fundamental experience, often conveyed in a sense of sound. Sometimes the two categories blended into written records that broke down barriers, as we shall see. I want to explore here the voices from across time for whom that sense of necessity, found expression. It might be as parody, comment or simply an observation and a quest to understand.

The beginning of our attempts to put the sound of the natural world into words, came from folk names, nick-names, and these were often based on description. The work of the oral historian George Ewart Evans, and the radio recordings of Denis Mitchell and Charles Parker in the mid-part of the twentieth century revealed that there was indeed a poetry of the vernacular in which a place can itself assume a poetic quality based on the sounds that belong to it, and are made in it, and this included above all, the voice. Ewart Evans and Parker were friends, one working in the Midlands and the other based in East Anglia. Of Parker's work, Ewart Evans wrote: 'Language in its origin and development is necessarily a *spoken* language and its strength is clearly seen in the way the common man uses it, particularly an old countryman who uses words and expressions that have survived the use of generations and the smoothing and polishing of thousands of tongues.'[3] It all begins in the air, and with air, as Aristotle had realized 300 years before the birth of Christ: 'That which sounds ... is that which produces motion in such air as is one in continuity with the hearing organ ... Nature uses the breath both for the internal warmth of the animal, which is a necessity, ... and for the voice.'[4] Across air comes language, or rather, the sounds that become language. We learn our native tongue at first like birds, by listening to others. Take a young nightingale and place it with birds of other species, and it will learn to sing like them. Put it back with its own kind, and it will adopt its familiar voice. So it is with human young, dialect and folk language. The imitative quality in sound, accent and dialect in bird and man, and the folk language that humans have adopted to name birds through the aural impression that they have made, link us as a species. Sometimes these names vary from region to region. A crake, for example, has been found to be a northern Britain name for a carrion crow, based on the hoarse sound it makes, but it is perhaps more generally associated – as the corncrake – with the landrail. In Gloucestershire, the lesser spotted woodpecker

[3] G. Ewart Evans, *Spoken History* (London: Faber and Faber, 1987), p. 150.
[4] Aristotle, *De Anima* (London: Penguin, 1986), pp. 177–9.

was sometimes known as 'the crank-bird', probably from its cry, resembling, as Kirke Swann, writing in 1913, suggested, 'the creaking produced by the turning of a windlass'.[5] Words can sometimes only reveal their meaning when spoken aloud in their original dialect. While the word 'cuckoo' is clearly onomatopoeic, in other languages we find various related names for the bird that have a similar sound quality: 'kuckuck' in German, 'coccys' in Greek, 'cuculus' in Latin. The general gaelic word is 'cuthag', while on the islands of Mull and Iona, it has been called 'cuach', once again, echoing the cry of the bird itself. There is a whisper of voices in sometimes hidden ornithological writings that reveal the sounds of birds, and our relationship with them, from Pliny the Elder to H. G. Adams's popular anthology of 1851, *Favourite Song Birds*, which collected romantic and often sentimental poetic reflections on birdsong that may well have proved a successful and cherished bedside book in many a Victorian household. The tone of this, typical of a number of works of the time, and in contrast to the more familiar and lasting texts discussed elsewhere, may be gathered from this extract from its 'Dedicatory Sonnet', to the famous soprano, Jenny Lind, known as 'The Swedish Nightingale', and presumably written by the compiler:

> Tis thine to wake the sympathetic chords
> That slumber in each heart not wholly dead;
> Thine is a power more eloquent than words;
> Thine is a soul that hath on music fed[6]

On a more cynical note, Thomas Hardy parodies the chattering classes in his socially satirical poem, 'The Spring Call' in which he apes the dialect and sometimes affected voices of various part of Britain, impersonating the blackbird; sounds, human and avian, living side by side. It also suggests to us that Hardy himself spoke with a Dorset dialect as well he might. The line, 'In Wessex accents marked as mine' in the first verse, gives us a clue:

> Down Wessex way, when spring's a-shine,
>   The blackbird's 'pret-ty de-urr!'
> In Wessex accents marked as mine
>   Is heard afar and near.
> He flutes it strong, as if in song

---

[5] H. K. Swann, *A Dictionary of English and Folk-Names of British Birds* (London: Read Books, 1913), p. 63.
[6] H. G. Adams, *Favourite Song Birds: Feathered Songsters of Britain*. 1851 (Reprinted Delhi: Pranava Books, 2021).

> No R's of feebler tone
> Than his appear in 'pretty dear,'
>   Have blackbirds ever known.
> Yet they pipe 'prattie deerh!' I glean,
>   Beneath a Scottish sky,
> And 'pehty de-aw!' amid the treen
>   Of Middlesex or nigh.
> While some folk say–perhaps in play –
>   Who know the Irish isle,
> 'Tis 'purrity dare!' in treeland there
>   When songsters would beguile.
> Well: I'll say what the listening birds
>   Say, hearing 'pret-ty de-urr!'
> However strangers sound such words,
>   That's how we sound them here.
> Yes, in this clime at pairing time,
>   As soon as eyes can see her
> At dawn of day, the proper way
>   To call is 'pret-ty de-urr!'[7]

The use of birdsong as a vocalized metaphor for human foibles and conditions represented a concept of sound as caricature and a societal and cultural shorthand that was available to all, whether they were equipped with scientific knowledge or not. If you could hear, you were aware of the natural world, and the freedom suggested by wild sound often acted as a poignant comparison with the confines of urban or rural poverty. Folk knowledge among the working classes was frequently a rich source for the expression of comment, but was circumscribed by the change in society brought about through the industrial development and the decline of farming labour, as machines took the place of workers. Oliver Goldsmith (1728–74) was a city-dweller, yet he could tune his voice to a note of melancholy as the rural world was eroded:

> No more thy glassy brook reflects the day,
> But choked with sedges works its weedy way;
> Along thy glades, a solitary guest,
> The hollow-sounding bittern guards its nest;
> Amid thy desert walks the lapwing flies,

---

[7] T. Hardy, *Collected Poems* (London: MacMillan, 1979), p. 244.

And tires their echoes with unvaried cries.
Sunk are thy bowers in shapeless ruin all,
And the low grass o'ertops the mouldering wall.[8]

Alongside established writers such as Goldsmith, and later Hardy, there are voices almost lost, that contain heartfelt responses to the sounds of nature, as if reflecting on the realization of a casting-out from innocence. All rose to express the sounds of their social classes, sometimes directly through dialect. At the same time, certain species of bird might be adopted to illustrate meanings and codes of thinking. Nancy M. Derbyshire has pointed to the fact that 'the labouring class poetic tradition came of age in the final decades of the eighteenth century, and with it the labouring-class bird … The works of late-century plebeian poets illustrate birds as prevalent symbols of labouring-class poetic sensibility and experience.'[9] In Britain such poets found empathy with the domestic robin, at once feisty and fragile, territorial and linked to year-round habitats, present through all weathers and climatic privations, and above all sociable. 'Robin' is a contraction of the Old English name, 'Robin Redbreast' and has links deep within English folklore, with roots in many superstitions, often associating the bird with the wren, possibly because of their relatively comparable size, as Kirke Swann illustrates:

The Robin and the Wren
    Are God Almighty's Cock and Hen:
    Him that harries their nest,
    shall his soul have rest.[10]

Perhaps more than any other bird, the British robin is linked to mortality. It has also been noted as a barometer, in the Kirke Swann's entry: 'It is a common Border belief that if the Robin sings from underneath a bush it will rain, but if he mounts to the top of a bush to sing, a fine day may be expected'. A Suffolk rhyme is:

If the Robin sings in the bush,
Then the weather will be coarse;
But if the Robin sings on the barn,
Then the weather will be warm.[11]

---

[8] O. Goldsmith, 'The Deserted Village' quoted in M. Drabble. In *A Writer's Britain*, M. Drabble (ed.) (London: Thames and Hudson, 1979), p. 69.
[9] N. M. Derbyshire, 'The Labouring-Class Bird' in *Birds in Eighteenth-Century Literature*, B. Carey, S. Greenfield and A. Milne (ed.) (Cham: Palgrave MacMillan, 2020), pp. 92–3.
[10] H. K. Swann, *Dictionary of English Birds*, pp. 198–9.
[11] Ibid.

The countryman or woman, tuned to the sounds around them, often had the ability to 'read' the signals of weather and seasonal change in similar ways to sailors at tides' turn and in veering storms.

A rich lore grew up around birdsong, and with it an empathy with the everyday presence of – in particular – the robin. After all, when the class structure turns against you, to whom do you relate as allies, other than your own people …and nature? Derbyshire draws our attention to the fact that in the eighteenth and early nineteenth centuries, 'labouring-class poets employed various strategies to foreground their social position in British society, including the revision of dominant poetic forms. Plebeian bird poetry was one strategy to cultivate sympathy and assert authorial independence.'[12] One striking example of this comes from the pen of William Lane (born 1744) from the village of Flackwell Heath in Buckinghamshire. His poem, 'On Reading a Poem Entitled "Modern Parnassus"' evokes not only the idea of a democratic literacy, but an implicit egalitarian view of society:

> But larks with soaring wing, and warbling throat
> Do not forbid the robbin's [sic] feeble note!
> The birds, tho' taught by nature to ascend,
> Do not alike their notes and wings extend;
> Tho' formed for flight, some round our hoses hop,
> And scarce dare ascend the chimney top;
> While some sublimely soar and reach a height,
> Beyond the outmost stretch of human sight.[13]

Also implicit in Lane's poem is the relationship between the height of flight and the sublimity of song; the closer to heaven, the more sacred and awe-inspiring the music. At the same time, the bird rooted to the everyday garden, field or hedgerow, sings a song that reflects the reality it experiences around it. Derbyshire quotes from the work of Robert Anderson (1770–1833) a calico printer from Cumberland, whose sonnet, 'To a Redbreast, Which Visited the Author Daily for Some Months', reflects both human emotion and social conditions by suggestion:

---

[12] N. M. Derbyshire, *Birds in Eighteenth-Century*, p. 107.
[13] Quoted in Derbyshire, p. 99.

> Emblem of Poverty! How hard thy fate
> When the wild tempests scowl along the sky!
> E'en now methinks thou wail'st thy absent mate,
> Singing thy love-lorn song: – just so do I.[14]

The shift from rural life to the industrial world becomes reflected in working-class poetry as the industrial revolution takes hold, with the migration from village to city, often enforced by the circumstances of poverty, dictating the need for survival. Edward Rushton (1756–1814) was a poet, writer and bookseller from Liverpool who began his life as a sailor on a slave ship, and as a result of what he witnessed, became an abolitionist, and well known as a writer of verse which espoused his views. After losing his sight, he founded a school for the blind, the first such in the world. In his 1806 poem, 'To a Redbreast in November', written near one of the Docks of Liverpool', he paints a clear picture of the robin as a spokesperson for the displaced worker or immigrant, surrounded by the noise, wealth and power of more confident and louder social classes:

> The Lark may reach the rose cloud,
> And strike his epic lyre aloud;
> The high-perch'd Throstle, clear and strong
> May roll his nervous ode along;
> The blackbird from the briery bower
> His deep tone'd elegy may pour[15]

Yet for all this, the humble robin, 'th' unbending minstrel', is not cowed or broken:

> All shivering in misfortune's storm,
>   While half nutrition wastes his form,
>   From fancy's heights beholds the crowd below,
>   And spite of varied ills, uncheck'd his raptures flow.[16]

The world of nature offered itself to rich and poor alike. The idea that the wealthy were driven to imprisoning birds in cages and aviaries, or even to blinding some to improve their song, gave meat to the concept of an indomitable voice uncowed by circumstance and oppression. While much was lost, as the countryman either worked in all weathers within a rural landscape, or sacrificed that world for dark

---

[14] Ibid., p. 97.
[15] Ibid., pp. 101–2.
[16] Ibid., p. 102.

streets, smoke and smog, there was yet a free spirit that could not be destroyed, and for these working-class poets, this was exemplified by birdsong and flight.

Beyond folklore and social comment, the rural writers of the seventeenth, eighteenth and nineteenth centuries still frequently celebrate the pure sounds of nature, notably as the industrial revolution began to impinge on the countryside in terms of structures and air quality. This work, in turn, ironically often became favoured by wealthy and literate city-dwellers as despatches from an idealized world, unspoilt and unsullied by affectation and so somehow preserving an essence that could be released at the turn of a page. Of this, there is no greater example than 'the peasant-poet' John Clare, who we shall discuss later. Likewise, Stephen Duck (c.1705–1756), born in the village of Charlton near Pewsey in Wiltshire into a poor-labouring family, was first by employment a thresher, mower and reaper, and then a poet who through his ability to express the rural muse, found himself transplanted from the sequestration of the Pewsey Vale to the Court of Queen Caroline, and the literary world of Pope and Swift. He became a poetic wonder of his time. The Queen listened to his poems at Windsor, awarded him an annuity, made him a Yeoman of the Guard and keeper of Duck Island in St. James's Park, had him educated for the Church, and in 1752 gave him the rectory at Byfleet in Surrey. Despite all the honour and recognition, life clearly ultimately became intolerable for Duck, and on 21 March 1756, he committed suicide by drowning himself in the River Kennet at Reading. What drove him to this, we do not know, but there may be something in the writing suggesting his relationship with the landscape, and the society within it, human and otherwise, was one of such total cultural immersion that he found himself incapable of existing when separated from it. His work often blends mankind and nature as one, finding like behaviour across species in poems such as 'The Thresher's Labour' where he draws an analogy between hedge sparrows and field workers, and in the process gives us a rare glimpse of the human sounds of rural life, on a hot day in the middle of the eighteenth century, at a meal break, fuelled one suspects, with certain amounts of cider and beer:

> The grass again is spread upon the Ground,
> Till not a vacant Place is to be found;
> And while the parching Sun-beams on it shine
> The Hay-makers have Time allow'd to dine.
> That soon despatch'd, they still sit on the Ground;
> And the brisk Chat, renew'd, afresh goes round.
> All talk at once; they seeming all to fear,

> That what they speak, the rest will hardly hear;
> Till by degrees so high their Notes they strain,
> A Stander-by can nought distinguish plain.[17]

The chatter grows louder and more animated; meanwhile, what he calls the 'tattling Croud' [sic] fails to notice that the weather is changing, until suddenly, caught in a sharp shower, 'their noisy prattle all at once is done, /And to the Hedge they soon for Shelter run.' Duck continues with his comparison between farm workers and the local bird population:

> Thus have I seen, on a bright Summer's Day
> On some Brake, a Flock of Sparrows play;
> From Twig to Twig, from Bush to Bush they fly;
> And with continued Chirping fill the Sky:
> But, on a sudden, if a Storm appears
> Their chirping Noise no longer dins our Ears;
> They fly for Shelter to the thickest Bush;
> There silent sit, and all at once is hush.

It is a touching sound picture of country life at the time and gives us a view through a window into history that somehow avoids an idealized pastoral vision, replacing it with a sense of human reality, written as it is with real affection and a sense of presence. This warm, good-humoured poem rings true, and it is not hard to understand how it opened a door for city-dwellers in the eighteenth century into other strata of their own time, while for us it records what has long fallen silent.

Such writing appeals because of its honesty; at the same time, there has always been a strain that referred back to an imagined 'golden age', usually beyond memory. Through into the twentieth century, an appetite has persisted for memoirs and poems from writers who had no literary aspirations, other than to capture a nostalgia for their own pasts. Many of these have found their way into print through small, independent and sometimes private publishers, intended often for little more than their own immediate community, while others have occasionally transcended that 'localness' and found a wider audience. Sometimes, this genre could be mawkish and oversentimental. Sometimes too, the bucolic has been consciously appropriated for manipulative political purposes; a notable

---

[17] S. Duck, 'The Thresher's Labour' in S. Carr (ed.) *Ode to the Countryside* (London: National Trust, 2010), pp. 46–7.

example of this came from the British Prime Minister, Stanley Baldwin, who, as late as the year of the English General Strike – 1926 – could use Golden Age nostalgia in the service of a manifesto for nationalism and patriotism, in an address called 'On England':

> The sounds of England, the tinkle of the hammer on the anvil in the country smithy, the corncrake on the dewy morning, the sound of the scythe against whetstone ... These things strike down into the very depths of our nature, and touch chords that go back to the beginning of time and the human race, but they are chords that in every year of our life sound a deeper note in our innermost being.[18]

Delivered during Baldwin's second term as Prime Minister, this notorious passage, evoking through sound, the idea of a time to which the nation should aspire again, came a year before his Conservative Government introduced the Trade Disputes and Trade Unions Act, 1927, to curb the power of the trade unions. The thinking behind it formed a mantra that would attract later politicians who could cynically use nostalgia, be it in words or in music to evoke a green and pleasant land, often in attempts to trigger nationalist fervour, from post-war restoration to Brexit. Pictures in the mind, be it individual or communal, are persuasive and lingering, and sounds from the past, even if they are sounds that exist in the imagination only, have the capacity to play a potent subconscious soundtrack. This ready market created a wealth of home-spun material of greatly varying quality, becoming an industry within the United Kingdom with the publication in 1977 of Edith Holden's posthumous *The Country Diary of an Edwardian Lady*, the facsimile of a 1906 hand-created text of sketches, notes and poems reflecting her rural life in the Midlands. It attained huge fame and sales, as well as many spin-off items of merchandise and design products for the home. It is perhaps notable that its major success was during the 1980s and 1990s, the Margaret Thatcher years, at a time of burgeoning wealth, business and social inequality, when, according to Thatcher's own words, there was 'no such thing as society.'[19] It is interesting also to note that even in the twenty-first century, there remains a perceived response to a kind of woolly pastoral that, at its most bland, has little or nothing to do with listening and looking, and more with a melancholy nostalgia for a life within a landscape that may or may not have existed, and which

---

[18] S. Baldwin, *On England, and Other Addresses* (London: Philip Allen, 1926), p. 7.
[19] M. Thatcher, in an interview in *Woman's Own* magazine, 1987, quoted *The Guardian*, 8 April 2013.

has been over time melded into the sunset glow of a persuasive semi-fiction, easily appropriated by an increasingly sophisticated commercial and political lobby. There is then the anthropomorphic element of response to birdsong and the visual context of rural landscape, encapsulating the sound as a shorthand linked in some strands of rural writing to the idea of a kind of 'Dream England.'

Then there is the sound of birdsong itself, existing before interpretation and layers of meaning, the human need to find a connection, be it direct or metaphorical. In some labourer-poems, the writers were not seeking to meet the natural world on its own terms, but to find sympathy through similarity for their conditions, advocacy to symbolize fragility and defiance in one object and a single voice. Complementary to this, there is the very nature of all sound: we hear it as it passes us, coming out of silence, peaking as our awareness grows and then fading back into silence or ambient noise. It is poignant because it is like us, present in the world only for a short time; it is untouchable, invisible and transient. Beyond and before this, however, there is its very existence, the pure sense of felt response that touches us *before* the brain tries to rationalize and attempt to explain, or anthropomorphise, and put us in the bird's position. The ear as microphone, the mind as receiver, the human senses as recording device, doing what electronics would later take on with increasing levels of (albeit unthinking) accuracy. In this poem by the Scottish writer, John Davidson, who lived from 1857 to 1909, are recorded the sounds of spring, the result one feels of active and conscious listening:

> About the flowerless land adventurous bees
>   Pickering hum; the rooks debate, divide,
>   With many a hoarse aside,
> In solemn conclave on the building trees;
> Larks in the skies and plough-boys o'er the leas
> Carol as if the winter never had been;
>   The very owl comes out to greet the sun;
>   Rivers high-hearted run;
> And hedges mantle with a flush of green.
> The curlew calls me where the salt-winds blow
>   His troubled note dwells mournfully and dies;
>   Then the long echo cries
> Deep in my heart. Ah, surely I must go!
> For there the tides, moon-haunted, ebb and flow;
> And there the seaboard murmurs resonant;

> The waves their interwoven fugue repeat
> And brooding surges beat
> A slow, melodious, continual chant.[20]

Almost every line in the poem contains a sound (It might perhaps be an interesting exercise for a sonic artist to use Davidson's poem as a template and create a sound work in which all the words were replaced by sounds, a soundscape poem orchestrated to the score of the poem). Although his life was shortened by his suspected suicide in 1909, Davidson lived to be respected and acknowledged as a writer by no less than W. B. Yeats, and he was to be an early influence on poets as diverse as Hugh MacDiarmid, Wallace Stevens and T. S. Eliot, whose sense of sound was to be the key in his ideas and expression of the auditory imagination. Even in this relatively straightforward sensory bucolic poem, phrases such as 'the interwoven fugue' of waves startle the ear and make their own sound in sympathy with the image they are depicting. We hear what John Davidson heard, but we do so through his filter. Here, the poet's consciousness does what a microphone cannot: it tunes itself mentally, focuses on the minutiae of the natural world to the exclusion of other sounds. In writing these verses, Davidson may be giving the impression of a spontaneous soundscape, happening simultaneously, the sonic equivalent of a slow panning shot in a film, taking in the various sounds in passing. It may be, however, that each sound reflected in the lines is the response of individual meditation, just as a microphone might isolate the item under study. Either way, the effect is the result of a series of decisions. In Ludwig Koch's birdsong recordings, he wanted us to listen to the example as just that – a sample, an artefact – and to do so, he was prepared to go to enormous lengths and pains, because he had to think for the equipment. Davidson does this by listening, reflecting and then seeking to find a way with words into a series of audio images that will make the reader 'hear' the original sound, as if present beside him, building cumulatively into a general wide-angle picture drawn from sonic particulars.

The late nineteenth century offered a ready market for populist writing aimed primarily at city dwellers, appearing in general magazines, journals and periodicals such as *London Society*, *The Cornhill Magazine* and the *St James Magazine* among many others. These outlets provided opportunities to gain a

---

[20] D. Davidson, 'Spring Song' in S. Carr (ed.) *Ode to the Countryside: Poems to Celebrate the British Landscape* (London: National Trust Books, 2010), p. 121.

readership for numerous now near-forgotten amateur and semi-professional authors with the capacity to reflect on and share experiences drawn from rural life, to feed the appetite, often, as we have seen, for a manufactured nostalgia, reflecting a world where the pace of life was purer and slower. In 1874, the Rev. J. G. Wood (1827–1889) drew a selection of his writings together from just such a variety of sources, into a volume published by Longmans, under the title, *Out of Doors*. Wood was for much of his life a priest in Oxford, although from 1854, he devoted himself increasingly to natural history writing, gaining a reputation as a popularizer of the subject. He accepted a readership at Christ Church, Oxford, and was an assistant chaplain to St. Bartholomew's Hospital, London from 1856 to 1862. From 1876, however, he devoted himself full-time to writing, settling from 1868 in Upper Norwood in South London, at the time a burgeoning suburb of new housing, attracting a growing generation of middle-class professionals and commuters. Of his many books, *Out of Doors* deserves some mention if only because it is quoted by Sir Arthur Conan Doyle in his Sherlock Holmes story, *The Adventure of the Lion's Mane*. These essays are full of focus for the leisure reader, an audience Wood understood well, and a flavour may be gained by this extract from 'A walk through a country lane', in which he trains the town-based reader to pause and listen: 'There goes the humble-bee, blundering along the flower-clad bank, with its steady, continuous drone, occasionally broken by a sharp, congratulatory buzz, as it alights on some untouched flower'.[21] For the most part, this is typical of a raft of semi-rural writing that created a sentimental picture of country life and the natural world that suited a certain element of Victorian sensibility. This same essay ends in similar vein:

> Calm and quiet is the evening now, the sounds of labour are hushed and the bright songs of happy birds are stilled in sleep. But Nature has her vespers as well as her orisons; the shrill cry of the bat and the deep humming of the circling beetle are psalms of praise as intelligible to sympathetic hearts as the sweet melody of feathered throats, or the pleasant sounds of busy insect wings.[22]

Full of sound indeed, but a soundscape that offers religious rather than social metaphors, and a world at peace with itself.

No matter what the motive for descriptive writing, when it comes to sound, we are listening to the writer listening, and in some cases, moralizing. At its best,

---

[21] J. G. Wood, *Out of Doors* (London: Longmans, Green, 1874), p. 83.
[22] Ibid., p. 114.

however, natural history writing involves moving partly towards another kind of being, learning how to inhabit a parallel entity (the sound itself) and then sharing a new perception, as it were, from the inside. As a reader, we must sometimes trade the ability to experience a reality for the capacity to rely on the expression and interpretation of others. In the last 200 years, humankind has created sounds in daily life that have exceeded those that issue from the natural world; prior to this, the sonic signals came more clearly, and there was a knowledge of how to read them that was instinctive. When Mozart wrote *presto*, the fastest sound-generator in his world would have been the galloping horse and the fastest object above him, a bird. The loudest sound for him might have come from the farmyard: the volume of a cockerel's call can reach a deafening 143 decibels, indeed so loud that when in full 'song', a quarter of the bird's ear canal closes to protect its own hearing. Gaius Plinius Secundus, Pliny the Elder (AD 23–79), claimed this was all done for humankind's benefit, 'which Nature designed for interrupting sleep and waking men to work … they prevent the sunrise from creeping up on us unnoticed, and announce the arrival of day with song and the song itself with the beating of their wings'.[23] A sense of relativity is useful, because scale is everything when considering – or reconsidering – the natural sound world. The sophistication of birds' use of sound can be extraordinary, and is almost infinitely variable, according to the voice in question. What sonically separates us – between species as well as between generations – is time. We live our sound lives through differing pitches and frequencies; in order to come close to understanding how bats and birds such as swifts, swallows and starlings negotiate the air, we need to slow the tape down. None of this was available to our listening ancestors. On the other hand, there are bird sounds that we can interpret with little difficulty, and mostly these are alarm calls: the scolding of a blackbird, or the shouting of magpies. Birds are aware of their own 'family' and they are aware of objects of danger, and because we too, as animals, are primed to grow alert in times of risk, we can read their cries of fear and warning more than any other sound they make. As to the *actual* sound, and its interpretation and imitation, that can be problematic. Writers have frequently turned away from seeking to do what the birds do better and have instead focused on the *feeling* the natural soundscape has upon the human listener. In the next chapter,

[23] Pliny the Elder, *Natural History* (London: Penguin Classics, 2004), pp. 143–4.

we shall hear how Chaucer's fourteenth-century poetic woods pulsed with the *idea* of the sounds he would have heard.

As we have observed, local knowledge and tradition often informed the naming of wildlife and plants, and in the case of birdlife, the most common recourse was through sound. The Norfolk name for a starling is a 'wheezer', which as Mark Cocker and Richard Mabey point out, 'nicely suggests the asthmatic breathy notes, while a low-pitched whistle in the song has been likened to "milk spurting from a cow's udder into a bucket".'[24] Like many other birds, starlings have historically sometimes been held captive, caged for the novelty of their powers of imitation. Today we have no need of this; we can hold the voice without imprisoning the bird, and the recording device is a humane and efficient cage in which to study the aural presence of the natural world. We are guilty of human arrogance, however, if we consider ourselves unique in our ability to listen and record. Hearing is necessary for survival in all species, and we know that some birds record sound mentally and retransmit it in their own sonic image. Starlings are well known to confuse birdwatchers with their imitations of not only fellow birds, but other creatures, and notably household objects; Cocker and Mabey cite 'a note published in 1952 of a starling that sounded so similar to a telephone bell that it had the author's wife running indoors to pick up the receiver. It was probably the first occasion that such mimicry was ever recorded.'[25] It was, however, certainly not the last, and starlings have continued to adapt to technology around them as it – and they – evolve. Likewise, the mynah bird has been known to impersonate any number of objects, including motorcycles and kettles, while at the top of the familiarity list for most would also be the parrot, particularly adept at mimicking the sounds of human language. An online blog from the Cornell Lab offers this explanation:

> Unlike songbirds, which produce sounds by vibrating membranes in two different syrinxes, parrots have only one syrinx, located at the bottom of the windpipe. This is somewhat similar to humans, who also have only one sound-producing organ, the larynx. Parrots also have long, muscular tongues that may be used in modifying sounds. Parrots also have forebrain areas involved in vocal learning and control of vocalisation that are not found in other birds.[26]

---

[24] M. Cocker and R. Mabey, *Flora Britannica* (London: Chatto and Windus, 2005), p. 429.
[25] Ibid., pp. 429–30.
[26] The Cornell Lab: https://www.allaboutbirds.org/news/which-birds-are-the-best-mimics/ (accessed 5 July 2022).

North America boasts several avian mimics, including mockingbirds, catbirds and thrashers, (the brown thrasher can perform something in the region of 2,000 different songs). Blue jays have been known to imitate several species of hawk. So, we are not the only critical listeners in the natural world; when we walk into our garden, there are many conscious presences attending to our every move and sound, and some of them may be learning how to impersonate us. As to meaning, this provides the age-old philosophical debate, and has long fuelled the poetry this book seeks to explore in terms of the nearness and farness of humankind and wildlife. As Elizabeth Eva Leach has written: 'Words, like rational-sounding pitches, can be deceptive when perceived solely by the ears; both could be the product of imitation. Ultimately, both are just particular types of sound: language too is just a sonic property.'[27] Leach goes on to quote Dante to illustrate the point:

> If it be claimed that, to this day, magpies and other birds do indeed speak, I say that this is not so; for their act is not speaking, but rather an imitation of the sound of the human voice – or it may be that they try to imitate us in so far as we make a noise, but not in so far as we speak. So that, if someone who said 'pica' [magpie] aloud the bird were to return the word 'pica,' this would only be a reproduction or imitation of the sound made by the person who uttered the word first.[28]

We may learn the *sound* of a foreign language, 'parrot-fashion', and perfect the 'music' sufficiently to be understood, but we will find it hard to comprehend the answer, unless it be through extra-linguistic qualities such as volume, hand signals and beyond that, passion. Thus, it works both ways; while the magpie, parrot or starling will not know the significance of the sound they copy, likewise, Basil Bunting once wrote a short and pithy poem which mocks our misunderstanding of a thrush's song. In the end, it is the pure music that we envy, and it is that which makes us write the poetry. On the other hand, if birds by any chance reflect on our presence, they seem, as far as we know, to lack the means or inclination, by and large, to express it. As Richard Smyth reminds us, 'no nightingale ever wrote an ode about John Keats and no snowy heron or Carolina parakeet ever painted a portrait of John James Audubon.'[29] In the

---

[27] E. E. Leach, *Sung Birds: Music, Nature and Poetry in the Later Middle Ages* (Ithaca and London: Cornell University Press), pp. 40–1.
[28] Dante, *De vulgari eloquentia*, pp. 4–5, quoted in Leach, p. 41.
[29] R. Smyth, *An Indifference of Birds* (Axminster: Uniform Books, 2020), p. 15.

next chapter, we shall venture into the green wood, and share the melancholy of minstrelsy. We may delight in the chorus of birds on a spring morning, but as the sixteenth century poet George Daniel (1562–1619) wrote of the robin, it is the indomitable voice, heard alone 'when the bleake face of winter spreads,' with which we identify, and from which we gain comfort:

> Poore Redbreast, carol out thy laye,
> And teach us mortals what to saye.
>   Here cease the Quire
> Of Ayerie Choristers; noe more
> Mingle your notes, but catch a store
>   From her Sweet Lire; [sic]
>   You are but weake,
> Mere summer Chanters; you have neither wing
> Nor voice in winter. Pretty Redbreast, Sing
>   What I would speake.[30]

Some would argue that birdsong is not song at all, while the French troubadours, who we shall meet shortly, would on the other hand have no doubt that it is. Whether or not the sound is understood, it is shared. This chapter began with the democracy of radio waves, and birds are like radio signals insofar as they can bridge national barriers without a passport, the same, or similar, sounds heard in countries thousands of miles apart. Their language – or song – is both untranslatable and universal. At its heart there remains a mystery, and because the reason and purpose – if there must always be such things – remain open to debate, meanings can be applied and appropriated. We have touched on a few voices from the field, some examples from many that hopefully form representative samples of first-hand experiences gained sometimes through amateur observation, and often as part of daily life. From these witnesses springs the idea that sounds from the natural world can be translated into new meaning within the mind, ideas that we may interpret as metaphors, an expression for the human condition. 'Sing/What I would speake', says the poet, in the hope that birdsong will do it better than he can. It becomes clear that common to both bird and humankind is orality.

---

[30] G. Daniel, 'The Robin' in S. Carr (ed.) *Ode to the Countryside* (London: National Trust, 2010), pp. 130–1.

5
Paths through the green wood

During the Middle Ages there was a perceived difference between natural sound, including birdsong, and human-created music. Elizabeth Eva Leach points out that 'the key feature that defines music [of that time] is its expression of a rationality, which human beings alone of all the sublunary animals also possess'.[1] On the other hand, long before that time, Pliny the Elder (AD 23–79) had written of the nightingale in his *Natural History*:

> There is the perfect knowledge of music present in one bird: the nightingale modulates its sound, at one time drawing out a long, sustained note with one continuous breath, now varying it by adjusting the breath, now making it staccato by checking it; or it suddenly lowers the note or, when it pleases, makes the sound soprano, mezzo or baritone. In short that little throat contains everything that human skill has devised in the complicated mechanism of the flute … The birds have several songs each; they are not all the same, and every bird sings its own repertoire.[2]

There were also writers who acknowledged meaning in sounds beyond human comprehension. In Sextus Empiricus's (*c*. second/ third AD) *Outlines of Pyrrhonism* the author acknowledges that 'even if we do not understand the utterances of the so-called irrational animals, still it is not improbable that they converse'.[3] Sextus and his thinking have a place in this argument because the Pyrrhonistic 'suspension of judgement', named for the philosopher Pyrrho of Elis (*c*. 360–270 BC) maintained that we are not capable of gaining a completely true knowledge of reality, and, for this reason, should refrain from making

---

[1] E. E. Leach, *Sung Birds: Music, Nature and Poetry in the Later Middle Ages* (Ithaca: Cornell University, 2007), p. 1.
[2] Pliny the Elder, trans. J. J. Healy, *Natural History: A Selection* (London: Penguin Books, 2004), pp. 145–6.
[3] Sextus Empiricus, trans. R. G. Bury, *Outlines of Pyrrhonism* (Lanham, MA: Prometheus Books, 1990), p. 38.

dogmatic and definitive statements abo--ut things we cannot fully understand. He supports this illustration of comprehension by pointing out that while we may not understand languages from other lands, this does not make them any less intelligible to their own kind, 'for in fact when we listen to the talk of barbarians we do not understand it, and it seems to us a kind of uniform chatter'.[4] He goes on:

> Whoever examines the matter carefully will find a great variety of utterance according to the different circumstances, so that, in consequence, the so-called irrational animals may justly be said to participate in external reason. But if they neither fall short of mankind in the accuracy of their perceptions, nor in external reason, nor yet (to go still further) in external reason, or speech, then they will deserve no less credence than ourselves in respect of their sense impressions.[5]

Birds, so far as our understanding tells us – and mostly it is male birds – 'sing' for two principle reasons: to declare territory and to advertise themselves to a prospective mate, but they make other sounds too, which can be almost as evocative – more so in particular circumstances. R. Murray Shafer has pointed out that avian sound is not always by any means vocalized:

> Birds may be distinguished by the sounds of their flight. The great slow clapping of the eagle's wing is different indeed from the tremulous shaking of the sparrow against the air ... The startled exodus of a flock of geese on a northern Canadian lake – a brilliant slapping of wings on water – is a sound as firmly imprinted in the mind of those who have heard it as any moment in Beethoven.[6]

Frequently, in literature, it is the place itself that is evoked, sound acting as a prompt to the imagination as part of a broader picture. Writers of the Middle Ages mostly stopped short of direct imitation, while using avian life as symbolic, metaphoric and sometimes anthropomorphic. Geoffrey Chaucer, (1340–1400) in his long poem, 'Troilus and Criseyde' lulls Criseyde to sleep to the anaesthetic singing of a nightingale in the green wood outside her window, evoking the scene thus:

> A nyghtyngale, upon a cedir grene,
> Under the chambre wal ther as she ley,

---

[4] Ibid.
[5] Ibid.
[6] S. R. Murray, *The Soundscape: Our Sonic Environment and the Tuning of the World* (Rochester: Destiny Books, 1994), p. 33.

> Ful loude song ayein the moone shene,
> Peraunter, in his briddes wise, a lay
> Of love, that made hire herte fressh and gay.
> That herkned she so longe in good entente,
> Til at the laste the dede slep hire hente.[7]

By way of an explanatory aside; I quote Chaucer in his original linguistic form here, because it seems to me that the *sound* of the language is a key part of the overall discussion of this book. Even a reader unfamiliar with early forms of English will, I hope, find a music and a logic of meaning when reading these texts aloud, rather than silently on the page. There is a wonderful prose poem of a lecture given by Gertrude Stein during her 1934–5 tour of America called 'What Is English Literature' [No question mark]. In it, there is this key passage, which not only stands as central to this point, but indeed, as a cornerstone for what this book is, at its heart, about. I quote it here as it is on her original page:

> You do remember Chaucer, even if you have not read him you do remember not how it looks but how it sounds, how simply it sounds as it sounds. That is to say because the words were there. They had not yet to be chosen, they had only as yet to be there just there. That makes a sound that gently sings that gently sounds but sounds as it sounds. It sounds as sounds that is to say as birds as well as words. And that is because the words are there, they are not chosen as words, they are already there. That is the way Chaucer sounds.[8]

Geoffrey Chaucer was born in London, the grandson and son of wine merchants. What was his environment – immediate and within close proximity? How did it sound then? These and other questions begin to interrogate our sense of where we are, and how our place in the chronological development of the landscape and soundscape fits with what was seen, heard and experienced before. In other words, context, situation and circumstance. Chaucer begins his most famous work, *The Canterbury Tales* with a Prologue, the opening of which sets a scene of burgeoning nature to which an imagined soundscape comes easily to the reading ear:

> Whan that Aprill with his shoures soote
> The droghte of March hath perced to the roote,

---

[7] G. Chaucer, 'Troilus and Crisyede' Book II In *The Complete Works of Geoffrey Chaucer* (Oxford: Oxford University Press, 1974), lines 918–24, p. 411.
[8] G. Stein, 'What Is English Literature', collected in *Look at Me Now and Here I Am* (London. Peter Owen, 2004), p. 42.

And bathed every veyne in swich licour
Of which vertu engendred is the flour;
Whan Zephirus eek with his sweete breeth
Inspired hath in every holt and heeth
The tendre croppes, and the yonge sonne
Hath in the Ram his halve cours yronne,
And smale foules maken melodye,
That slepen al the nyght with open ye
(So priketh hem nature in hir corages);
Thanne longen folk to goon on pilgrimages[9]

In his paper, 'Chaucer's Nightingales', Marvin Mudrick draws attention to the specificity of the birds that play their sonic part in this landscape/ soundscape, the spring song that awakens the heart and the migratory urge, in particular pointing to these three lines:

And smale foules maken melodye,
That slepen al the nyght with open ye
(So priketh hem nature in hir corages)

'Three lines of description' as Mudrick states, 'that set apart the kind, note its habits and impulses, and establish its remoteness from the purposes of man, while the weightless fluidity of the first line affirms without taking into custody the beauty (free, alien, uncoerced) of bird-song'.[10] For all his powers of description evoking the natural world, direct imitation of birds' voices are rare in Chaucer, even in his *Parlement of Foules*, with this notable exception:

The goos, the cokkow, and the doke also
So cryede, 'Kek Kek! Kokkow! quek quek!' hye,
That thurgh myne eres the noyse wente tho.[11]

Chaucer's poem is an imaginative account of the celebration of Valentine's Day among the birds. Interestingly, the lines quoted above contain the only phonetically rendered bird imitation in the whole long poem. For the most part, the birds use human language; here only does rank and file heckling, amounting to a rabble of noise, interrupt the debate. The idea for the poem stemmed from

---

[9] G. Chaucer in 'General Prologue to The Canterbury Tales' in ibid., p. 17.
[10] M. Mudrick, 'Chaucer's Nightingales'. *Hudson Review*, vol. 10, no. 1, 1957, pp. 88–95, https://doi.org/10.2307/3847575.
[11] G. Chaucer, from 'The Parlement of Foules', in *Chaucer.* p. 316, stanza 72.

a folklore story that birds married for one year, and the fancy behind the poem is that these marriages all occur on the same day, and Chaucer makes this Valentine's Day. Subsequent research has shown that, as so often, there is an element of fact behind the legend. Charles Darwin, in his *The Descent of Man*, illustrates this in relation to the common magpie, which 'used to assemble from all parts of Delamere Forest, in order to celebrate the "great magpie marriage."'

> They then had the habit of assembling very early in the spring at particular spots, where they could be seen in flocks, chattering, sometimes fighting, bustling and flying about the trees. The whole affair was evidently considered by the birds as one of the highest importance. Shortly after the meeting they separated, and were then observed by Mr Fox [Darwin's informant on the phenomenon] and others to be paired for the season.[12]

From this Darwin goes on to draw broader conclusions: 'We may conclude that the courtship of birds is often a prolonged, delicate and troublesome affair. There is even reason to suspect, improbable as this will at first appear, that some males and females of the same species, inhabiting the same district, do not always please each other, and consequently do not pair.'

Chaucer's poem is thought to have been written for Richard II and his Queen, Anne of Bohemia, effectively making this the first literary celebration of Valentine's Day as a festival for lovers. In this allegorical work, it is mostly the birds that take centre-stage, but Chaucer does not by any means neglect the broader soundscape. In stanzas twenty eight and twenty nine, he offers fourteen lines of exquisite sound evocation, in which the forest becomes a heaven-like place of dappled light and whispering sound:

> A gardyn saw I ful of blosmy bowes
> Upon a river, in a grene mede,
> There as swetnesse everemore inow is,
> With floures white, blewe, yelwe, and rede,
> And colde welle-stremes, nothyng dede,
> That swymmen ful of smale fishes lighte,
> With fynnes rede and skales sylver bryghte.
>
> On every bow the bryddes herde I synge,
> With voys of aungel in here armonye;
> Some besyede hem here bryddes forth to brynge;

---

[12] C. Darwin, *The Descent of Man* (London: Penguin Books, 2004), p. 457.

> The litel conyes to here pley gonne hye;
> And ferther al aboute I gan aspye
> The dredful ro, the buk, the hert and hynde,
> Squyrels, and bestes smale of gentil kynde.
>
> Of instruments of strenges in acord
> Herde I so pleye a ravyshyng swetnesse,
> That God, that makere is of al and lord,
> Ne herde nevere beter, as I gesse.
> Therwith a wynd, unnethe it myghte be lesse
> Made in the leves grene a noyse softe
> Acordant to the foules song alofte.[13]

The humour in the poem – and there is much – stems often from the conflict between the hierarchy within the bird kingdom, satirising as it does the attitudes and behaviour of human social classes; the noble birds of prey support the idea of mating within the tradition of courtly love, while for others, such as the goose and duck, the idea of pining for an unrequited love is a ludicrous waste of time: why not cut out all that nonsense and simply get on with things! So with this in mind, there is a parody of current conventions of chivalry and romantic etiquette. Hence the interjection: 'Kek Kek! quek quek!', voices from the floor of the debate, making a derisory rumpus in order to debunk their pretentious fellow birds, who have such elevated views of the accepted romantic social standards of courtship.

Approximately 250 years after Chaucer's extended Valentine's Day card, John Donne wrote 'An Epithalamion' for the marriage of Princess Elizabeth, the only daughter of James I, to Frederick, Elector Palatine, which was celebrated on 14 February 1613. Whether or not Donne had been inspired by *The Parlement of Foules*, his poem begins with two stanzas positively vibrating with birdsong, and a clear reference at the very start of the significance of the day for birds of the marrying kind:

> Hail Bishop Valentine, whose day this is,
>   All the Aire is thy Diocis,
>   And all the chirping Choristers
> And other birds are thy Parishioners.
>   Thou marryest every yeare
> The Lirique Larke, and the grave whispering Dove ... (& c.)[14]

---

[13] *The Parlement of Foules*, pp. 312–13, stanzas 28–9.
[14] J. Donne, *The Complete Poems of John Donne* (London: Dent, 1985), pp. 192–3.

Like Donne, Geoffrey Chaucer moved in prestigious and learned circles, and he spent his career closely involved with the royal court circle of England. By the time of writing *The Parlement of Foules*, he was well travelled; he had been sent on court business at various times to France, Flanders, Spain and Italy, more than once to Genoa, and it is likely that he had a good working knowledge of languages, and would have absorbed a number of current literary forms such as rhyme royal, of which he was the first practitioner in English. His travels will have also made him aware of, and influenced by, texts from across Europe, in particular, the work of the troubadours, who flourished at the height of the Middle Ages in southern France. Despite the work of some eminent supporters such as Ezra Pound and H. J. Chaytor, the myth persists of the troubadour as a figure akin to Alan-a-Dale, a legendary minstrel who first appeared in the seventeenth-century broadside ballad, 'Robin Hood and Allan-a-Dale'. It was perpetuated in some of the poems of Sir Walter Scott, notably 'The Troubadour' and 'The Lay of the Last Minstrel', and thence found its way into romantic fiction and Hollywood movies. In reality, the troubadours wrote poetry in the Provençal language between 1095 and 1295, works created under the patronage of noblemen of the time, and thus aimed primarily at courtly and aristocratic audiences. Although growing out of the South of France in an area that has become known by scholars as Occitania, comprising Aquitaine, Languedoc, Auverne and Provence itself, the culture quickly spread south into Spain, and east to Italy, and in its influence, further afield yet. The importance of troubadour poetry is hard to over-state; as Anthony Bonner has written: 'It was the first lyric poetry in any modern language, and all other lyric poetry in Europe either descends from it or was at one time tremendously influenced by it.'[15] Their songs of romantic love have been an inspiration from Dante to Verlaine and Rimbaud, to modern popular music lyricists.

In order to fully comprehend their sound world, we must picture a landscape full of forests and rich in animal species. A troubadour's expression of the landscape was a sophisticated one, in contrast to much of what had previously been written, strongly embedded in the sounds and colours of the natural world. Indeed, the Provençal troubadours such as Bernart de Ventadorn showed a highly developed idea of birdsong in their lyrics, including what one might

---

[15] A. Bonner, *Songs of the Troubadours* (New York: Schocken, 1972), p. 1.

almost characterize as an envy born of realization that it was more poetic than their own songs. Take, for example, this, from one of Ventadorn's songs:

> When I perceive the skylark lift
> His joyful wings to dawn's new light
> Then let himself, unthinking, drift
> For his sense of joy, his heart's delight,
> Ah then an envy fills my soul
> To see the ecstasy that others find[16] (& c.)

The word 'troubadour' means finder or maker; the twelfth- and thirteenth-century troubadours of southern France were indeed the makers, or composers of songs, largely writing for the courtly entertainment of feudal nobles and their ladies of the region. Makers they were, but for the most part, they were not the singers of their songs, a task they left to specialist 'joglars',[17] known for their sweet voices, and conduits for the troubadours' inspiration, typically expressions of pure or refined love: *fin'amor*. The nobles for whom the troubadours wrote, lived in locations where the world of nature pressed against their castle walls and the use in a number of the songs of this world and its inhabitants as protagonists and sometimes as metaphors, is not an infrequent device, as in this opening to Jaufre Rudel's 'The Nightingale':

> Deep in the woods, the nightingale
> Knows of love, and pleads love's prize,
> And as he speaks his joyful tale
> Yet to his love he turns his eyes.
> Streams run clear, green pastures rise.
> And as the spring's new pleasures start,
> Joys grow and spread inside my heart.[18] (& c.)

In sound terms used here, there is a close-up on the bird, then a wider panning perspective of the stream and pastureland spreading into the distance. The verse is like an establishing scene for the drama that follows over six stanzas, a blueprint for a detailed sonic picture and in its own right, a soundscape full of rustling spring-like sound. Likewise, another prolific troubadour, Peire d'Alvernhe, active in the mid-twelfth century, at his best could show 'a flawless joining of

---

[16] B. D. Ventadorn, 'The Skylark'. English extract in my version.
[17] 'Joglar' is an old Provençal word, from which derives 'jongleur', an itinerant minstrel.
[18] J. Rudel, 'The Nightingale.' English extract in my version.

sound, rhythm, meaning and association'.[19] Here, in 'Nightingale, for me Take Flight' the songbird is a messenger, and therefore capable of both music, and by implication, speech, equipped to fulfil the role of emissary:

> Nightingale, on my behalf take flight,
> Swiftly to my lady's home repair;
> There speak to her of my lonely plight,
> Learn for me how in truth she fares[20]

We find this tradition continuing in the work of the Welsh poet, Dafydd ap Gwilym (c.1320–50), a major interpreter of sound in the natural world from the time (although time and place in the life of Dafydd are somewhat hazy). There are a few datable events in his poems which would lead us to believe him to be active in the mid-fourteenth century, putting a possible birthdate at around 1320: in other words, a near contemporary of Chaucer. As to a birthplace, tradition has it as Brogynin, a hamlet near Aberystwyth. Again, clues in the poems suggest that this part of Wales was more familiar than elsewhere, and there is no indication that he travelled beyond the boundaries of his native country. As to his themes, there is much in common with the troubadour tradition: 'Love and Nature are the prime subjects of his poetry, and the two are very frequently blended, [and he represents] his love-theme most characteristically in an idealized woodland setting, in which he imagines himself in a *deildy* or house of leaves and branches in which to shelter with his chosen sweetheart'.[21] Yet over and above the love themes, it is the forest itself – its ambience, moods and presence – that occupies and inspires him to often breathtakingly expressive flights of description. Even the wind itself has a voice, capable of delivering a message to the beloved, and thus there is a common understood language between human and non-human that binds the two into a single, inseparable whole. As H. Idris Bell has suggested, 'usually natural description, if introduced at all, serves ostensibly as no more than a background for the love interest; but often here, we feel that love was but an excuse for a delineation of nature'.[22] It is right that we should speak of his work as a link between the troubadours and the Welsh bardic forms. On the other hand:

---

[19] A. Bonner, *Songs of the Troubadours* (New York: Schocken Books, 1972), p. 69.
[20] P. D'Alvernhe, 'Nightingale for Me Take Flight.' English extract in my version.
[21] R. Bromwich. In D. A. Gwilym (ed.), *Selected Poems* (London: Penguin, 1985), p. xiv.
[22] H. I. Bell, *Dafydd Ap Gwilym: Fifty Poems* (London: The Honourable Society of Cymmrodorion, 1942), p. 27.

If Provence is certainly the ultimate source of much that we find in Dafydd, it is equally certain that the influence was not direct. Not only is it in the last degree improbable that Dafydd could have read a line of the Provençal poets, but the differences between his work and theirs are too great to make direct borrowing credible. The influence of Provence was, however, ubiquitous in the Middle Ages; and the poetry of Ovid, [b. 43 BC] to which the Troubadours were indebted, was transmitted to the world of medieval poetry through other channels also, notably the songs of the *clerici vagantes*.[23] [24]

As for Dafydd, however, his main influences are to be found principally in the song thrush, the nightingale and the skylark, although his cast of characters spreads beyond avian life into hunted animals and indeed, the woods themselves. Everything is as one, sacred and supported by an intimate knowledge. 'The singing of both woodlark and sedge-warbler which frequently sing loudly on summer nights has often been taken for that of the nightingale. Dafydd, though, would seem too keen an observer to have made such a mistake. Exact and accurate observation, and in the case of the skylark and seagull, an almost mystical reverence pass into the world of fantasy.'[25] He knows through the instinctive experience of the countryman rather than from scientific research or second-hand reading that, for instance, the tit has a darting flight and weak cry, that the nightingale builds a deep nest; he has watched the behaviour of the magpie in the nesting season, has seen the salmon leap the weir, and draws this information from observation rather than reading. The skylark and the song thrush are evoked in verse of luminous quality, while in 'The Woodland Mass' ('Offeren y Llwyn') Dafydd brings together the thrush and the nightingale to celebrate mass in the depths of the forest as though they are ordained priests. Any translation is bound to diminish the original, and there is no way an English version can do justice to the turns of phrase, linguistic complexities and nuances within the verse. What follows, therefore, is intended as an indication only of the spirit of Dafydd's original poem. It is morning, and from the start, the location is clearly imaged, with its effect on the poet:

---

[23] *Clerici Vagantes*: A medieval Latin term meaning 'wandering clergy', applied in early canon law to a vagrant itinerant clergy leading a wandering life due either to the fact that they had no beneficence or that they had deserted the church in which they had been attached. A certain type of *vagantes* began to appear in France during the twelfth century who were also minstrels. They became masters of poetic form, and were later known as 'goliardi', another term for this travelling European clergy writing satirical Latin poetry from the twelfth/thirteenth centuries.
[24] Bell in *Fifty Poems*, p. 16.
[25] Ibid., p. xxiii.

> I was in a pleasant place today,
> beneath mantles of green-leafed hazels,
> listening as day broke,
> to the consummate song of the cock thrush,
> performing a shining *englyn*,
> full of portents and pure lessons[26]

In the final stanza we celebrate communion with the birds, in the green and gold chapel of the forest:

> I heard in clear bright language
> a long and ceaseless recitation
> through strong, voluble language
> his reading of the gospel to the throng
> with no unseemly haste or pompous pause.
> Raised on an ash-hill,
> with leaf as consecrated wafer-bread,
> the slim and eloquent nightingale,
> poet of the vales and streams,
> ringing her Sanctus bell with a clear trill,
> lifting the Host to the sky above the grove,
> chanting adoration to God the Father
> with a chalice full of ecstasy and love.[27] (& c.)

We know how things have changed, how great forests have died or been felled, how sea levels have risen and engulfed landscapes, how deserts have replaced verdant woods and grasslands, so in order to 'hear' the places as audiences and readers from the past heard them, we must exercise our imagination. Yet the work of these writers: Chaucer, the Troubadours and Dafydd ap Gwilym, speak to us of a time that is vibrant with sound, landscapes that seem to rustle and quiver with unseen life, and which offer, frequently by suggestion, an auditory presence that plays like a soundscape, a wild track field recording behind the action of the stories of courtly love, longing and sometimes rumbustious living that remain vivid through their word pictures. More, they demonstrate the ear and mind as faithful and accurate recording devices, supported by knowledge and the power of emotional and poetic responses that continue to convey with precision, the sound world of the Middle Ages.

---

[26] '*Englyn*'. A traditional Welsh and Cornish short poem form.
[27] Gwilym, 'The Woodland Mass'. My version.

To find the earliest attempts at *direct* representation of bird sound, that is to say of the sort briefly attempted in Chaucer's *Parlement of Foules*, we must go much further back in time. One of the first texts to make extensive textual/sonic use of birds in drama must be *The Birds* by Aristophanes (*c.* 446–386 BC), which won the second prize in the Athens City Dionysia of 414 BC. The story follows Pisthetaerus, a middle-aged Athenian who persuades the world's birds to create a new city in the sky (thereby gaining control over all communications between men and gods), and is himself eventually miraculously transformed into a god-like bird figure himself, replacing Zeus as the pre-eminent power in the cosmos. It is a work remarkable for its mimicry of birds, and Aristophanes was a skilful enough writer to interpret birdsong such as the nightingale both in textual sound-form and choral speeches of reflective lyricism:

> Loveliest of birds,
> Flaxen nightingale,
> The voice of all birdsong,
> Your tremulous flute
> Makes you princess of the woods.
> You are here, you are here,
> Emerging from your secrecy,
> From where you accompanied
> Our chorusing. Now you have come,
> Adored – oh darling bird!
> Open our voices once again,
> Pure with your silver song.[28]

The gender of the nightingale may in reality be doubtful, given that it is the male that sings, but the hymn to the ethereal nature of sound is vivid. There are other speeches in which the chorus (made up of twenty-four different species of bird) copies directly into words that may be read on the page more as stage directions open to interpretation in production than as literal intentions for sound reproduction. For example:

> Just so the swans,
> Tio-tio-tio – tiotinx,
> The swans on Hebrus' banks
> With slowly beating wing

---

[28] Aristophanes, *The Birds* [Parabasis: 676–800]. My version.

Tio-tio-tio – tiotinx,
Sang praise, Apollo's praise,
Tio-tio-tio – Tiotinx
To the heavens beyond the clouds.[29]

The key factor in exploring sound representation in dramatic works is a somewhat obvious one: it is precisely that they were designed not to be read from the page, but performed on stage, and in this instance, as part of the Greek theatrical tradition, across vast open spaces. When the word is spoken, pitched or sung from the page or on a stage by an actor or actors, rather than absorbed through the eye, a 'recording' of impressions is shared or transmitted, moving us a step closer to the idea of dissemination through broadcast. We bring words into meaning within our heads both silently and noisily at the same time. This takes us back to the birth of poetry, and its original orality, as in the singing of Homer and the performance of bardic ballads as performed by troubadours. It is to do with aural signals which, while they may convey sounds to the imagination silently, are actually written to be heard aloud through the medium of the human voice. These voices may quite often possibly be specific, known to an audience of the time, as today we may interpret amusing lines of dialogue through the mannerisms of a particular actor or stand-up comedian. For example, William Shakespeare (1564–1616), working nearly two thousand years after Aristophanes, in a more controllable theatrical environment, was developing his plays with a company in which there were men and boys with particular vocal characteristics who were familiar to their theatre-going public, actors who possessed qualities that could be exploited in terms of pitch, range, timbre and dexterity. The question of the conveyance of sound through the voice in Shakespeare is interestingly discussed by Francis Berry in his book, *Poetry and the Physical Voice*:

> Did writing or voices outside himself ... enable Shakespeare, as a poet, to go beyond himself? To orally convey by means of his surrogates areas of experience which otherwise (if he had limited himself simply to his own resources) he could not have conveyed, and possibly could not therefore have conceived? If so, did this distension, by employing the vocal instruments of others, cause him to abandon after 1603 (or by whatever year the sequence of *Sonnets* was concluded) compositions which he conceived in terms of delivery through his own voice?

[29] Ibid.

Can we discern or re-hear the vocal history of one or more of these surrogates behind the printed signs on the pages – the pages of the earliest printed texts with all their peculiarities of punctuation, lineation, spelling, capitalisation?[30]

The voice in which words are spoken is a conduit that aids interpretation.

Supplemented by the added values a stage performance can bring – those special elements such as acoustics, scenery, costumes, lighting and so on – it can and should be transformative. The visuals complement the sound. Psychologically, because hearing is only one of the senses, working in unison with the others, we listen to things differently according to what we are seeing at the time, and one voice can change that experience as opposed to another, much as the interpretation of a particular musician will offer nuances of meaning that through a soloist with other ideas will be subjected to different emphases. It is a melancholy fact that the nature of performance prior to recording was ephemeral, so if a text was written for performers with certain vocal characteristics, they are lost to us, while we may yet have the written text. The human voice, like birdsong, is, as the French writer Pierre de la Primaudaye (1546–1620) wrote, 'invisible to the eyes, so it hath no body wherby [sic] the hands may take hold of it, but is insensible to all the senses, except the hearing … [and so] … vanisheth away suddenly.'[31] We can therefore, but, do our best to gain an idea of the sonic landscape in which images were opened by such voices, and to do so, we should pause and consider where to turn to evoke the rural environment in which Shakespeare grew up and lived, and within which many of his plays were either set or inspired:

> When Phoebus lifts his head out of the Winter's wave,
> No sooner doth the earth her flowery bosom brave,
> At such time as the year brings on the pleasant Spring,
> But hunts-up to the Morn the feath'red Sylvans sing[32]

The words are not by Shakespeare but by his Warwickshire contemporary and compatriot, Michael Drayton (1563–1631) taken from his *Poly-Olbion*, published in two parts in 1612 and 1622. Drayton was born nineteen years before Shakespeare, and died fifteen years after him. *Poly-Olbion* is a topographical poem describing England and Wales, and the section on Drayton's home county is full of the birdsong to which Shakespeare would have grown up hearing; it is a

---

[30] F. Berry, *Poetry and the Physical Voice* (London: Routledge and Kegan Paul, 1962), pp. 121–2.
[31] P. De la Primaudaye, *The French Academie*, quoted in J. Richards, *Voice and Books in the English Renaissance* (Oxford: Oxford University Press, 2019), p. 39.
[32] M. Drayton, *Polyolbion* (London: John Russell Smith, 1876), thirteenth song, p. 145, lines 41–5.

regular poetic catalogue. He details the linnet, woodlark, wren, goldfinch, jay and bullfinch and many others, as well as placing them all in the auditory landscape, evoking strong pictorial images in the mind. Thus, we cannot visit Shakespeare's natural sound world without embedding it in Drayton's ringing soundscape. Although he lived in London, he was born at Hartshill, west of Nuneaton, and is known to have returned occasionally to the Warwickshire village of Clifford Chambers, just a couple of miles away from Stratford. Shakespeare and Drayton are thought to have been friends, and we know that Drayton worked with Philip Henslowe and the company called the Admiral's Men on about twenty plays between 1597 and 1602, so he was very much part of the same professional world, for the most part living in London, but revisiting his Warwickshire home from time to time. It was during this period that Drayton and Shakespeare would have been in proximity to one another and may have met. There is a colourful story that links Drayton indirectly to Shakespeare's death: According to John Ward, writing in the 1660s, Shakespeare had a 'merrie meeting' with Drayton and the playwright Ben Jonson in 1616, at which Shakespeare apparently 'drank too hard' and 'died of a feavour [sic] there contracted'.[33]

Be that as it may, if we wish to regain a sense of how that part of England sounded, and felt to be in at that time, then Drayton is helpful and his topographical writing may conceivably have provided inspiration for Shakespeare. Particularly in *Poly-Olbion* Drayton specialized in the description of landscape features. The poem is divided into thirty songs, written in alexandrine couplets, consisting in total of almost 15,000 lines of verse. Drayton intended to compose a further part to cover Scotland, but no section of this work is thought to have survived. Each song describes between one and three counties, exploring their topography, traditions and histories, and copies were illustrated with maps of each county, drawn by William Hole. It is in 'The Thirteenth Song' that Drayton's tour of Britain reaches Warwickshire, beginning thus:

> Brave Warwick; that abroad so long advanc'd her Bear,
> By her illustrious Earls renownéd everywhere;
> Above her neighbouring Shires which always bore her head.
> My native Country then, which so brave spirits hast bred[34]

---

[33] John Ward, once vicar of Stratford-on-Avon, quoted in *Birmingham History Forum*, 7 November 2010: https://birminghamhistory.co.ukeforum/index.php?threads/poly-olbion-c1620-by-drayton.33828/ (accessed 23 May 2022).

[34] M. Drayton, *Polyolbion, Thirteenth Song*, p. 146.

The Forest of Arden, the scenic world of *As You Like It*, is remembered in a time when it was richer, wider and deeper than in Drayton's, and therefore Shakespeare's:

> Muse, first of Arden tell, whose footsteps yet are found
> In her rough wood-lands more than any ground
> That mighty Arden held even in her height of pride;
> Her one hand touching Trent, the other Severne's side.
> The very sound of these, the Wood-Nymphs doth awake:
> When thus of her own self the ancient Forest spake[35]

Yet there is enough left to provide a platform for the sounds, magically evoked by Drayton, of a Warwickshire dawn chorus:

> And in the lower grove, as on the rising knole,
> Upon the highest spray of every mounting pole,
> Those Quiristers are perch'd with many a speckled breast.
> Then from her burnish'd gate the goodly glitt'ring East
> Gilds ever lofty top, which late the humorous Night
> Bespangled had with pearl, to please the Morning's sight:
> On which the mirthful Quires, with their clear open throats,
> Unto the joyous Morn so strain their warbling notes,
> That hills and valleys ring, and even the echoing air
> Seems all compos'd of sounds, about them everywhere.[36]

He then goes on to give the spotlight to some of the avian stars: 'The Throstell, with shrill sharps', 'The Woosell near at hand, that hath a golden bill/ As Nature had him mark'd of purpose', the Merle, 'Upon his dulcet pipe' and 'the Nightingale hard-by,/In such lamenting strains the joyful hours doth ply/As though the other birds she to her tunes would draw',[37] this last picking up the idea from Aristophanes that the nightingale was the voice of them all. Drayton's is a landscape in which birdsong was heard and known, a world understood as not something separate from daily living, but a surround-sound in which rural life was lived. Place, time and circumstance are all part of the sound world, as much as the sounds themselves, and they were known at first hand to Drayton and to Shakespeare, while to the city-dweller unfamiliar with the rustling whisper of

---

[35] M. Drayton, *Polyolbion*, p. 144, lines 13–18.
[36] Ibid., lines 44–4.
[37] Ibid., lines 55–5.

woodland and the voices of its inhabitants, Portia in *The Merchant of Venice* has a word or two:

> The crow doth sing as sweetly as the lark
> When neither is attended; and, I think,
> The nightingale, if she should sing by day,
> When every goose is cackling, would be thought
> No better a musician than the wren.
> How many things by season season'd are
> To their right praise and true perfection![38]

Shakespeare elsewhere in his work also mentions the blackbird (ousel-cock), bunting, chough, cock, cormorant, cuckoo, daw, dive-dapper, dove, duck, eagle, falcon, finch, fowl, guinea hen, hedge sparrow, heron, jay, kestrel, kingfisher, kite, lapwing, loon, magpie, mallard, martin (martlet) osprey, ostrich, owl, paraquito, parrot, partridge, peacock, pelican, pheasant, phoenix, pigeon, popinjay, quail, raven, rook, sea gull, snipe, sparrow, starling, swallow, swan, thrush, turkey, vulture and woodcock. Through these references, the sense of sound becomes either relevant or significant in various ways; for instance, birds and the animal kingdom can be metaphors, just as can the broader elements of storms and other weather conditions. Sound-wise, the mournful sound of the dove is invoked in *A Winter's Tale*, when Paulina says at the play's end:

> I, an old turtle,
> Will wing me to some wither'd bough, and there
> My mate, that's never to be found again,
> Lament till I am lost.[39]

A woodland can be an uplifting place, carolled by Amiens (clearly a troubadour at heart) in *As You Like It*:

> Under the greenwood tree,
> Who loves to lie with me,
> And tune his merry note
> Unto the sweet bird's throat[40]

---

[38] W. Shakespeare, *The Merchant of Venice*, act V, scene i.
[39] W. Shakespeare, *The Winter's Tale*, act V, scene iii.
[40] W. Shakespeare, *As You Like It*, act II, scene v.

On the other hand, it may be a place of retreat, of contemplation and melancholy, a parallel world in which to escape the noise and turmoil of the modern world, as Valentine considers in *The Two Gentlemen of Verona*:

> This shadowy desert, unfrequented woods,
> I better brook than flourishing peopled towns:
> Here can I sit alone, unseen of any,
> And to the nightingale's complaining notes
> Tune my distresses and my woes.[41]

That bird again! Nevertheless, these are above all locations where the place is part of the action. Duke Senior, in *As You Like It* 'Finds tongues in trees, books in the running brooks, /Sermons in stones, and good in everything'.[42] Meanwhile, Tamora, in *Titus Andronicus* comes to a natural environment of a very different kind:

> A barren detested vale you see it is;
> The trees, though summer, yet forlorn and lean,
> Overcome with moss and baleful mistletoe;
> Here never shines the sun, here nothing breeds,
> Unless the nightly owl or fatal raven[43]

When it comes to the birds and animals that inhabit these places, it is not so much their sounds as their actions with which Shakespeare frequently makes his point; sometimes on the other hand, the two go together with great potency, whether it be as escape:

> As wild geese that the creeping fowler eye,
> Or russet-pated choughs, many in sort,
> Rising and cawing at the gun's report,
> Sever themselves, and madly sweep the sky,
> So at his sight away his fellows fly[44]

Or as attack:

> The eagle suffers little birds to sing,
> And is not careful what they mean thereby,
> Knowing that with the shadow of his wings

---

[41] W. Shakespeare, *Two Gentlemen of Verona*, act V, scene iv.
[42] W. Shakespeare, *As You Like It*, act II, scene i.
[43] W. Shakespeare, *Titus Andronicus*, act II, scene iii.
[44] W. Shakespeare, *A Midsummer Night's Dream*, act III, scene ii.

> He can at pleasure stint their melody[45]

Or as portent:

> The owl shrieked at thy birth, an evil sign;
> The night-crow cried, aboding luckless time;
> Dogs howled, and hideous tempest shook down trees;
> The raven rooked her on the chimney's top,
> And chattering pies in dismal discords sung.[46]

We share this world. The sounds and actions of the natural environment belong to all life across time, and while we may make metaphors, at the end, we are all, even subconsciously, straining for sounds we either understand or do not; we listen, and interpret what we hear in terms of beauty or fear, beneficent gift or potential threat. In this we become observers, outsiders, while we struggle to be much more than that. Tom McFaul points to Caliban in Shakespeare's last play, *The Tempest*, and that character's facility for listening and interpreting: a blend of fear and wonder. 'His ability to *hear* is crucial, and may suggest an unfallen aspect to his character, for it was thought that the ability to hear the music of the spheres was something lost in the Fall of man. The sadness with which these lines end embodies the sadness of being a conscious human.'[47]

> Be not afeard, the isle is full of noises,
> Sounds, and sweet airs, that give delight and hurt not.
> Sometimes a thousand twangling instruments
> Will hum about mine ears; and sometimes voices,
> That, if I then had waked after long sleep,
> Will make me sleep again; and then, in dreaming,
> The clouds, methought, would open and show riches
> Ready to drop upon me; that when I waked,
> I cried to dream again.[48]

For the rest of us, we are indeed fallen, cast-out beings. Birds may represent angels, be they loving or vengeful, but what we hear in the green wood may be less of a threat than what lies beyond it, even if it is sometimes harder to understand and interpret.

---

[45] W. Shakespeare, *Titus Andronicus*, act IV, scene iv.
[46] W. Shakespeare, *Henry VI, Part 3*, act V, scene vi.
[47] T. McFaul, *Shakespeare and the Natural World* (Cambridge: Cambridge University Press, 2015), p. 184.
[48] W. Shakespeare, *The Tempest*, act III, scene ii.

6

Science in Arcadia: The road to Selborne

As with dance or abstract painting, pure sound can act on our senses to open doors to individual worlds of imagination, each of them intensely personal and direct. Working with our other faculties, it can create interactions and juxtapositions in the mind that are the very stuff of poetry. Poetic drama as we saw in the previous chapter, may have the potency of a text absorbed in silent contemplation, while providing the visual and auditory opportunities of performance. Sound is above all, the fundamental medium of storytelling, as witness Shakespeare's words in *King Lear*, a play which revolves crucially around the relationship between seeing and not seeing, imaginary perceptions and reality. Near the end of the play, Lear has this exchange with the blinded Gloucester:

> **Lear** *You see how the world goes.*
> **Glos** *I see it feelingly.*
> **Lear** *What, art mad? A man may see how this world goes, with no eyes. Look with thine ears.*[1]

In 1711, Alexander Pope (1688–1744) anonymously published his long poem, *An Essay on Criticism*. Within it, there are fourteen lines which have become known as 'Sound and Sense'. As with the rest of the work, it is written in couplets, and employs a number of auditory devices, including assonance and sibilance. Above all, however, it is a poem that offers advice on how a writer should set up his or her poetic 'recording equipment'. The section begins:

> True ease in writing comes from art, not chance,
> As those move easiest who have learn'd to dance.[2]

He proceeds to set out the crucial relationship between style and content:

---

[1] W. Shakespeare, *King Lear*, act 4, scene vi.
[2] A. Pope, 'An Essay on Criticism' in *Collected Poems* (London: Dent, 1969), lines 362–3, p. 67.

> 'Tis not enough no harshness gives offence,
> The sound must seem an echo of the sense:
> Soft is the strain when Zephyr gently blows,
> And smooth stream in smoother numbers flows;
> But when loud billows lash the sounding shore,
> The hoarse, rough verse should like the torrent roar.[3]

Pope was writing in response to examples of work he was reading at that time, some of which he saw as the bland equivalent of comfortable country estate landscape painting. 'Sound and Sense' may be applied as advice to any recordist of the natural world who uses words rather than a microphone, and indeed, the audio producer on location 300 years later, still seeks to capture the startling moment of reality, to surprise the ear and engage the imagination, and above all, to avoid the obvious, and eschew the cliché. At the heart of it all, it is about listening: active and critical listening, but with a view to sensing the meaning behind the sound itself, rather than sound for its own sake. Content first, then presentation in an appropriate style; Pope makes this point in some lines immediately before:

> But most by numbers judge a poet's song:
> And smooth or rough, with them, is right or wrong:
> In the bright muse, though thousand charms conspire,
> Her voice is all these tuneful fools admire;
> Who haunt Parnassus but to please their ear,
> Not mend their minds; as some to church repair,
> Not for the doctrine, but the music there.[4]

The seventeenth and eighteenth centuries brought fundamental changes in the world: the incursions of the industrial revolution on one hand, and what we know of as The Age of Enlightenment, the development of ideas based on the pursuit of knowledge by means of reason, and the evidence of the senses, on the other. There were ground breaking works such as René Descartes' *Discourse on Method* in 1637 ('Cogito, ergo sum – I think, therefore I am') and Isaac Newton's *Principia Mathematica* in 1687. By the start of the nineteenth century, the world would be in some ways transformed, in other ways, not. Certainly new questions were being asked, and to some degree answered, and fresh modes of business

---

[3] Ibid., lines 364–9.
[4] Ibid., lines 337–43, p. 66.

and industry were emerging, but within a climate of apparent certitudes, came many fresh uncertainties and issues. As Pope applied method and technique to poetry, so the world and its contents were interrogated in terms of why and how? The Royal Society was founded in 1660 with the motto, 'Nullius in verba – Take Nobody's word for it'. This new thinking applied to all forms of science, not least our exploration and questioning of the particularities and circumstances of the natural world. In the introduction to his vast *Dictionary of Birds*, written between 1893 and 1896, Alfred Newton looked back on the period, and was able to state confidently that 'the foundation of scientific ornithology was laid by the joint labours of Francis Willughby and John Ray'.[5] Willughby (1635–72) and Ray (1627–1705) met when the former was a student of the latter at Trinity College, Cambridge. They became lifelong friends, and were between them, with others such as the clergyman and natural philosopher John Wilkins (1614–72), advocates of a new way of studying science, relying on observation and classification, rather than the received authority of Aristotle and the Bible. We need to place them here, midway in the story, both chronologically and culturally, because it is from their thinking that the literature and sound awareness of ornithology changed and began its development into fresh and original modes of expression. Willughby lived a short life, just thirty-six years, but in that time he established a body of work that was to be a revelatory foundation for future studies. He was a founding member of The Royal Society, and pioneered the finessing and differentiation of bird studies through identification of their distinguishing features in ways that greatly influenced approaches to natural history. This quality, with Ray's involvement, was refined to include features of observation not previously undertaken, including sound, as in this extract from the *Ornithology*, a description of the Redshank, which includes the following: 'The legs long and red, the ungues black; it hath the posticus [hallux or fourth toe]; these she stretches backwards in flying which make amends for the shortness of the tail; it makes a piping noise'.[6] Willughby was engaged on his monumental *Ornithology* when he died in 1672, and the work was posthumously completed by Ray, whose lifelong and greatest interest was botany. While he accepted the task as a debt to the dead, and while also making full use of Willughby's researches, he was himself working on a series of books, each of which marked a new epoch in

---

[5] A. Newton, *Dictionary of Birds* (London: Elibron Classics, 1866), p. 7.
[6] F. Willughby, *The Ornithology of Francis Willughby. Book II* (Reprinted Delhi: Pranava Books, 2018), p. 19.

their particular field. Ray was a left field observer with an acute ear, and he was, as well as focusing on his work on botany and ornithology, a keen observer of the sounds of human language; he published *A Collection of English Words*, a work he returned to and expanded at various times in his life. The first edition, published in 1674, shortly after Willughby's death, was later further developed for a new version in 1691, which became the starting point for a further study of dialect and folk speech. Both editions of the *Collection* divide language into northern and southern variants, and we may imagine that Ray's highly tuned ear extended its attentions into the avian world.

A large part of the Willughby/Ray studies involved the minute examination of birds; they collected, shot or were sent as specimens, dissected and analysed every part of their anatomy, and the *Ornithology* catalogues, chapter by chapter, the findings, species by species. This does not dwell a great deal on sound, (dead birds do not sing), but from time to time there are references, often in relation to human response, as with rooks, who 'build together upon high trees about gentlemen's houses, who are much delighted by the noise they make in breeding time.'[7] Ray was first a botanist, but as his biographer, Charles E. Raven, has pointed out, he belonged to a time when aspects of science and indeed the arts, were not as compartmentalized as they subsequently became:

> The astonishing feature of his [Ray's] career is not his mastery of a single subject, but the range of his knowledge ... In these days of specialisation it is difficult to believe that a man could make himself expert in the whole of zoology literally as a sideshow and in the intervals of his main study; and Ray himself never claimed to have done so. But the fact remains that after Willughby's death he set himself to produce books on birds, fishes, mammals and reptiles, and insects; and that these books, even more than his botanical writings, laid the foundation for serious scientific progress in each subject.[8]

Without the work of Willughby and Ray, the interpretation of sound in the natural world within the broader context of the minutiae of scientific observation would be hollow and incomplete. At the same time, celebration of the natural world and its soundscapes burgeoned alongside the detail of their scientific examination. A factor in hearing sounds across landscapes in the sixteenth, seventeenth and eighteenth centuries relied, notwithstanding the growth of industrial noise,

---

[7] Ibid., p. 124.
[8] C. E. Raven, *John Ray, Naturalist: His Life and Works* (Cambridge: Cambridge University Press, 1950), p. 308.

on the relative *absence* of sounds that we have subsequently imposed on the environment. Sounds carried further through the landscape because there were less competing obstacles in the way, enabling localized soundscapes to be heard, read and 'recorded'. It is the minutiae of sound environments that enabled writers such as Izaak Walton and Gilbert White to focus on the local and communicate those truths to a wider audience. This is the real world of the naturalist-listener, beginning in the particular, and relating implications to the global, communicating silently to the reader's mind. We are not only surrounded by sound, in a real sense, we ARE sound, insofar as our thoughts, memories and reactions manifest themselves within our head in a kind of silent ambience that plays continually, even during sleep. So a text can suggest the idea of sounds that, while being physically unheard, make themselves felt in the mind through recognition, memory or pure imagination. Izaak Walton, Gilbert White and later Charles Darwin and the nineteenth century nature writer, Richard Jefferies, heard what was immediately around them, albeit it through the filters of the physical and the psychic and for their readers, this was part of the magic and discovery of their soundscapes.

Izaak Walton, (1593–1683) came as an outsider to the rural world, a native of Stafford who published *The Compleat Angler* in 1653, over twenty years before Willughby/Ray's great book, but through his characters, there is conveyed knowledge leavened by the joy of the landscape and the outdoor life. Walton's work spoke vividly to city dwellers, and those who sought rural havens as recreation. He himself lived for many years in London, where he kept an ironmongery shop in Fleet Street. John Donne was one among many distinguished friends. In an 1935 essay on Walton, the novelist John Buchan wrote of his *Angler* in terms of its value as a 'recording', a preservation of a way of life and recreation, written from a particular standpoint, and with direct kinship to White's *Selborne*: 'It is a transcript of old English life, a study of the folk heart…It unfolds the heart and soul of the angler – not necessarily the sportsman, but the *angler* – a man who sees nature through the glass of culture, the townsman and the gentleman'.[9] The book is structured as a dialogue between a hunter (Venator) and a fisherman (Piscator, the author), while later editions added a third character in the form of a falconer, whom Walton names 'Auceps', acknowledging the very air as a key part of the story:

---

[9] J. Buchan, Introduction to I. Walton. *The Compleat Angler* (Oxford: Oxford University Press, 1974), pp. xxii–xxiii.

Auceps: The element that I profess to trade in, the worth of it ... is of such necessity that no creature whatsoever, not only those numerous creatures that feed on the face of the earth, but those various creatures that have their dwelling within the waters, every creature that hath life in its nostrils, stands in need of my element.[10]

The natural world is wonderful for its own sake and in its own right, Walton declares, God has placed it there for the benefit of Mankind. Turning his attention to avian life, Auceps continues with a vivid introductory passage on the benefits of birds to humankind:

The very birds of the air ... are both so many, and so useful and pleasant to mankind, that I must not let them pass without some observations. They both feed and refresh him; feed him with their choice bodies, and refresh him with their Heavenly voices ... those little nimble musicians of the air, that warble forth their curious ditties, with which Nature hath furnished them to the shame of Art.[11]

Of course, by no means everyone had the opportunity or the financial means to invest time in rural recreation and leisure. Poverty was a hindrance to the indulgence of travel, and even in the country, the rural poor frequently did not have the energy or inclination to contemplate the delights Walton and his mercantile and business colleagues enjoyed. There were also urban and suburban dwellers for whom the natural world was best enjoyed through the filtered and sometimes through idealized texts of books and journals, just as even today there are monthly magazines trading in dream landscapes and country house living as escapism. Nevertheless, Walton's is a highly tuned, specialist book, reflecting on an aspect of the world from a particular perspective in a way that White was to do in his Selborne garden and John Bunyan did in *The Pilgrim's Progress*. With this in mind, let us return to Auceps, and his celebration of birdsong:

At first the lark, when she means to rejoice, to chear [sic] her self and those that hear her, she then quits the earth, and sings as she ascends higher into the air, and having ended her Heavenly imployment [sic], grows then mute and sad to think she must descend to the dull earth, which she would not touch, but for necessity ... But the nightingale ... breathes such sweet loud musick out of

---

[10] I. Walton, *The Compleat Angler* (Oxford: Oxford University Press, 1974), p. 25.
[11] Ibid., p. 26.

her little instrumental throat, that it might make mankind to think Miracles are not ceased. He that at midnight (when every labourer sleeps securely) should hear (as I have very often) the clear airs, the sweet descants, the natural rising and falling, the doubling and redoubling of her voice, might well be lifted above earth and say: 'Lord, what Musick has thou provided for the Saints in Heaven, when thou affordest bad men such musick on Earth!'[12]

The description of the sound itself is precise, and the idea that something as perfect as this must be sacred, links Walton's thought to the idea of Holy Nature, propounded in the poetry of Dafydd ap Gwilym, discussed in the previous chapter. Further, it is interesting to note that for Walton, something so pure MUST be feminine in origin; no male of any species could ever create such sublime sound! So, side-by-side, we see how the observed sound of the countryside reflected social, cultural and scientific perspective and geographic location; and the two were interrelated. Writers recorded what they heard, but each within their own agenda and place.

As if against a faint rumble from beyond the surrounding hills, the naturalists and scientists continued the minutiae of their observations, while the poets and lyrical thinkers sought to preserve the spirit of things. The naturalist and illustrator Eleazar Albin (1690–1742) was born seven years after the death of Izaak Walton, possibly in Germany, although little is known of his early life. By 1701 he claimed to be in Jamaica, but certainly we know he was in England by 1708, because by that time he was married and living in Piccadilly, London. Considered to be among the greatest of entomological book illustrators of the eighteenth century, he created works on insects, fish, spiders and two on birds, sometimes working with his daughter, Elizabeth Albin, including, in 1738, *A Natural History of English Songbirds*. This is an indexed catalogue, each illustration accompanied by quite detailed text, demonstrating his ability as an observer and naturalist in addition to that of an artist. Albin's writing reveals a love of the natural world, and in particular avian life, a gift to ornament the world for the human species, as witnessed in his introduction:

> Singing birds are so pleasant a part of the creation, whether we consider their variety, beauty or harmony, that the animal world does not afford more agreeable objects to the eyes, no, none that so sweetly gratifies the sense of hearing. They were undoubtedly designed by the Great Author of Nature on purpose to

[12] Walton, pp. 26–7.

entertain and delight Mankind, who, for the generality, are well pleased with these pretty innocent creatures.[13]

Albin's is a handbook for identification, education and celebration; it is also a guide for caging, keeping and feeding of birds in the home, both as entertainment and investment. The starling, for example, whilst 'it has a wild, screaming uncouth note, yet for its aptness in imitating Man's voice, and speaking articulately, and his learning to whistle diverse tunes, is highly valued … and when well taught, will sell for a great deal of money, five guineas or more.'[14] (Mozart, buying his famous starling as a pet in 1784, would have testified that it was money well spent.) Albin's interest in keeping wildlife may be considered to be professional as much as recreational; he describes his illustrations as 'exactly copied from Nature' and with his daughter, who undertook much of the colouring of the illustrations, the value of his work and its chief 'selling point' in his view was the fact that they were pictured from life, 'which is wanting in the books that have hitherto been published on this subject'.[15] We may therefore deduce that this close observation also informed his observation of sound. As the title of the *Natural History* requires, Albin gives clear and often somewhat florid descriptions of birdsong, in addition to offering discussions relating to markings, habits of nest building and seasonal behaviour. Here, for example, is part of his description of the sound of the blackbird:

> He … is one of the first that proclaims the welcome spring, by his shrill harmonious voice, as if he were the harbinger of Nature, to awaken the rest of the feathered Tribe to prepare for the approaching season: and by the sweet modulation of his tuneful accents, endeavours to delight the hen, and allure her to submit to his embraces, even before there are leaves on the trees, and whilst the frosts are in the fields.[16]

Izaak Walton himself, one suspects, would have enjoyed this description of the bird's voice ringing out on a clear, bright winter's morning. He was a quietist, a listener who approached the rural landscape and its inhabitants with an awareness bred of being a non-resident. He understood that when we encounter unfamiliar environments or places that are beyond our daily habitual activities, we absorb our surroundings differently; in effect, we become more like infants

---

[13] E. Albin, *A Natural History of English Songbirds* (London: R. Ware, 1738), p. i.
[14] Ibid., p. 11.
[15] Ibid., p. ii.
[16] Ibid., pp. 1–2.

in terms of how we absorb information. The new-born child is paying REAL attention to a huge number of new stimuli at the same time, absorbing things that to an adult would be a distraction, due to the capacity for focusing on the requirements of living as we grow, or would simply go unnoticed, due to familiarity. A city-dweller sensitive to environment, such as Walton, would encounter rural sound – or lack of it – as a dramatic contrast to the world of his daily life and work. The actress, poet, novelist and dramatist Mary Robinson (1757–1800) gives us a sound-sense of London in about 1774, a century after Walton's time, in her 'London's Summer Morning' answering any mistaken thought that the soundscapes of urban Britain offered a more contemplative listening experience than today:

> The din of hackney-coaches, waggons, carts;
> While tinmen's shops, and noisy trunk-makers,
> Knife-grinders, coopers, squeaking cork-cutters,
> Fruit-barrows, and the hunger-giving cries
> Of vegetable-vendors, fill the air.[17]

In just over forty lines, Robinson gives us a vivid portrayal of metropolitan cacophony, ending with the suggestion that any writer would find literary ideas awaiting expression, driven from the mind by the city's noise, giving rise to the poem she has now created in their place: 'and the poor poet wakes from busy dreams,/To paint the summer morning'.[18] William Hazlitt, in one of his essays, was to recreate a sound picture of London a few decades later, as we shall see in the next chapter. While differing in detail, it observed a similar effect on the nerves. In contrast to this, the comparative imagined peace of the rural world offered a country idyll and escape, which found a ready urban market and willing audience, and was to increase as the towns and cities changed through the Industrial Revolution. It was also of course for the most part an idealized dream for those who lived at a distance from the reality, although subject to correctives by works by writers with first-hand knowledge, such as 'The Farmer's Boy', a poem by Robert Bloomfield (1766–1823) which offers what we may imagine to be a farmyard sound picture that reads like a rural counterpart to Robinson's city soundscape, hardly a scene of bucolic calm:

---

[17] M. Robinson, *The Longman Anthology of Poetry* (London: Longman, 2006), p. 596.
[18] Ibid., p. 597.

> The clatt'ring Dairy-Maid immers'd in steam,
> Singing and scrubbing midst her milk and cream,
> Bawls out 'Go fetch the cows!' – he hears no more;
> For pigs, and ducks, and turkeys throng the door,
> And sitting hens, for constant war prepar'd;
> A concert strange to that which late he heard.[19]

In some ways, the division of centuries can be misleading; culturally and socially, we might consider there to be 'long' seventeenth, eighteenth and nineteenth centuries, in so far as literary and artistic trends tended to 'bleed' over such arbitrary boundaries of time, affected as they were by other criteria. Among those 'other criteria' we may consider sound as being an increasing presence; it is human nature to notice change, and rural sound changed increasingly from the late seventeenth century onwards. Some writers were aware of it – or at least incorporated it into their work – more than others. We see poets and other writers, influenced by the work of contemporaries in one era, growing to fruition in another, as in the case of Anne Finch, Countess of Winchelsea (1661–1720), who wrote her first works in the 1680s, while a poem by her in praise of the nightingale's song was created in 1713, and indeed, in its yearning, seems to hearken back further, to some of the envy in the songs of the troubadours:

> Exert thy voice, sweet harbinger of spring!
> This moment is thy time to sing,
> This moment I attend to praise,
> And set my numbers to thy lays.
> Free as thine shall be my song,
> As thy music, short or long[20]

The dream of idyllic rural sound, largely the province of the wealthy and leisured landowning classes, became increasingly confronted by a new reality as science and industry progressed. Anne Finch, nee Kingsmill, was at one time a part of the court of James II, where she met and married Heneage Finch, Gentleman of the Bedchamber to the Duke of York. They fell from grace when James fled in 1688, and left London and court life to settle in 1690 at Eastwell Park in Kent. Through this time, Anne continued to write, and was a friend and

---

[19] R. Bloomfield, 'The Farmer's Boy' in S. Carr (ed.) *Ode to the Countryside* (London: National Trust, 2010), p. 44.
[20] A. Finch, 'To the Nightingale' in R. Lonsdale (ed.) *Eighteenth Century Women Poets* (Oxford: Oxford University Press, 1989), p. 13.

correspondent of Pope. The beginning of the eighteenth century saw a continued flourishing of the pastoral vision of landscape, celebrated in landscape painting, often commissioned by wealthy estate owners to show off their manicured and sculpted lands, much of it acquired through wealth gained from the dubious practices of the British Empire, colonization, social inequality and enslavement. The classical idea of nymphs and shepherds, personified by Corydon, Daphnis and Ganymede in ancient pastoral poems and fables tended to cast a glow over images of rural life in literature. In 1709, Anne Finch was writing such poems as 'A Pastoral Dialogue between Two Shepherdesses', whom she named Dorinda and Silvia, perhaps conceived within the groves of Eastwell Park as she gazed over the lake and landscaped grounds of her adopted home:

> Silvia pretty nymph! Within this shade
> Whilst the flocks to rest are laid,
> Whilst the world dissolves in heat,
> Take this cool and flowery seat,
> And with pleasing talk awhile
> Let us two the time beguile.[21]

Against this artificiality, as the eighteenth century progressed, a particular strand of writing containing a new social awareness, gradually began to emerge, and as it developed, this broke the peace and tranquillity of the idyll. New meanings were sought; for some they were metaphysical, for others, social, and this complexity became an increasing factor in poetic soundscapes as the Industrial Revolution brought mechanization to the countryside, and the Age of Reason gained pace. Among the most vociferous of the balancing writers against social injustice and bucolic fiction was the Suffolk poet, George Crabbe (1754–1832). From the start of his two-book poem, *The Village*, Crabbe attacked the outdated and outmoded image of the rural life as it was reflected in literature:

> Fled are those times, when, in harmonious strains,
> The rustic poet praised his native plains:
> No shepherds now, in smooth alternative verse,
> Their country's beauty or their nymphs rehearse;
> Yet still for these we frame our tender strain,
> Still in our lays fond Corydons complain,
> And shepherds' boys their amorous pains reveal,

---

[21] Ibid., pp. 9–10.

The only pains, alas! they never feel.[22]

Village life, as Crabbe saw it in 1783, was absorbing the ills of the urban: 'Here, in disguise, the city's vice is seen' and returning to his birthplace of Aldburgh as curate in 1781, saw – and heard – at first hand, the effects of alcohol as closing time at the local pubs disturbed the peace:

> And hark! the riots of the Green begin,
> That sprang at first from yonder noisy inn;
> What time the weekly pay was vanish'd all,
> And the slow hostess scored the threat'ning wall;
> What time they ask'd, their friendly feast to close,
> A final cup, and that will make them foes;
> When blows ensue that break the arm of toil,
> And rustic battle ends the boobies' broil.[23]

Between this, and Crabbe's most famous sequence, *The Borough*, there is clear development of some of the sound themes, turning to the specifics of archetypal personalities, and the social tragedies some of them represented, among them Abel Keene, a village schoolmaster who is led astray and finally hangs himself, and Peter Grimes, known mostly today as the basis for Benjamin Britten's opera of the same name. The work takes the form of a series of 'letters' written in heroic couplets like *The Village*, and its realism conveys character and place with remarkable power. When characters converse, it is with a real voice, and we hear them through the lines speaking from their own points of view: '"What is a church?" "A flock," our Vicar cries, / "Whom bishops govern and whom priests advise"' while on the other hand, '"What is a church?" – Our honest Sexton tells, / "'Tis a tall building, with a tower and bells …"'[24] Amongst the murmur of voices, there are sound pictures that give us the place almost as if Crabbe is showing us around the village, or playing us a location-based sound file. The opening book ends with an image of Aldburgh at evening:

> When various voices, in the dying day,
> Hum in our walks, and greet us in our way;
> When tavern-lights flit on from room to room,
> And guide the tippling sailor staggering home;

---

[22] G. Crabbe, 'The Village' (Book 1) in *the Poetical Works of George Crabbe* (Edinburgh: Gall and Inglis, 1854), p. 17.
[23] Ibid. (Book Two), p. 24.
[24] Ibid., 'The Borough', letter II, pp. 96–7.

There as we pass, the jingling bells betray
How business rises with the closing day:
Now walking silent, by the river's side,
The ear perceives the rippling of the tide;
Our measured cadence of the lads who tow
Some enter'd hoy,[25] to fix her in her row;
Or hollow sound, which from the parish-bell
To some departed spirit bids farewell![26]

Crabbe gives us the place as he saw and heard it, and indeed as we, given the appropriate moment and situation, might hear it still. That said, location, time, climate and circumstance change the acoustics of the individual's field of consciousness, and thus their immediate interpretation of the world around them, just as an echoing church or a soundproofed studio alters the effect of voices and musical instruments. A bird singing on a city lamp post sounds differently to the same bird in the depths of woodland.

Meanwhile, outside the fields of social comment and satire, poets during the eighteenth century began to learn how to sing with a new and growing awareness of the natural world informed by the works of Willughby and Ray and the development of scientific knowledge. James Thomson (1700–1748), whose most famous work, *The Seasons* was to be such an inspiration to John Clare, acknowledged the coming together of art and science in his poem in memory of Isaac Newton: 'Th'aerial flow of sound was known to him, /From whence it first in wavy circles breaks' he wrote, and 'The noiseless tide of time, all bearing down/To vast eternity's unbounded sea,/Where the green islands of the happy shine,/He stemmed alone'.[27] The Scottish-born Thomson wrote his memorial to Newton in 1727, (Newton had died in March of that year) and it was during this time that he was working on *The Seasons*, as well as his first play, *Sophonisba*, completed in 1730. In the final lines of 'To the Memory of Isaac Newton', Thomson accepts the challenge of science to poets as interpreters of the new knowledge, finding common ground in which to celebrate jointly the workings of a beneficent creator, and Newton himself as, like himself, a servant in the cause of understanding:

---

[25] An anchor hoy: A vessel equipped for raising or handling anchors and chains.
[26] G. Crabbe, 'The Borough' Book I (Edinburgh: Gall & Inglis, 1854), p. 96.
[27] J. Thomson, 'To the Memory of Sir Isaac Newton' in R. Lonsdale (ed.), *The New Oxford Book of Eighteenth-Century Verse* (Oxford: Oxford University Press, 2009), p. 189.

> But who can number up his labours? Who
> His high discoveries sing? When but a few
> Of the deep-studying race can stretch their minds
> To what he knew: in fancy's lighter thought,
> How shall the Muse then grasp the mighty theme?[28]

Meantime, while absorbing the thinking of scientists, poets were tuning their listening and textual instruments to reflect with a new acuity the sonic presences of the natural world. In his 'Ode on the Spring', written in 1742, Thomas Gray (1716–71) hears in a pause between rural labour, the minutiae of life:

> Still is the toiling hand of Care;
> The panting herds repose.
> Yet hark, how through the peopled air
> The busy murmur glows!
> The insect youth are on the wing,
> Eager to taste the honeyed spring.[29]

Gray's ear was true and observant, pulling sound out of the air as Walton's angler pulled a fish from the stream. Insect life plays its sonic role too in his most famous poem, the 'Elegy Written in a Country Churchyard', in which, from the start, he sets the scene through the relationship between sound and light:

> The curfew tolls the knell of parting day,
> The lowing herd wind slowly o'er the lea,
> The ploughman homeward plods his weary way,
> And leaves the world to darkness and to me.
> Now fades the glimmering landscape on the sight,
> And all the air a solemn stillness holds,
> Save where the beetle wheels his droning flight,
> And drowsy tinklings lull the distant folds;
> Save that from yonder ivy-mantled tow'r
> The moping owl does to the moon complain
> Of such as, wand'ring near her secret bow'r,
> Molest her ancient solitary reign.[30]

---

[28] Ibid., pp. 190–1.
[29] T. Gray, *Lonsdale*, p. 349.
[30] Ibid., pp. 354–5.

It is still a pastoral world, tinged with gothic melancholy, but underpinned by close, active listening. Thus Gray, Thomson, Pope in his *Pastorals*, and William Cowper, notably in his long poem, *The Task*, recorded the natural world with an increasingly informed and detailed sense of aural observation. And science and the rural dreamscape of English downland combined too in *The Natural History of Selborne* by the Rev Gilbert White, who wrote in a letter on 18 April 1768, a note that in its very detail underlines the stillness of his 'sequestered vale': 'The grasshopper-lark began his sibilous note in my fields last Saturday. Nothing can be more amusing than the whisper of this little bird, which seems to be close by though at a hundred yards distance; and when close at your ear, is scarce louder than when a great way off.'[31] White explored bird language not in terms of human response, but as a debate between bird and bird. It isn't *meant* for humans to understand after all. To give it its full title, *The Natural History and Antiquities of Selborne*, takes the form of a series of letters to two fellow naturalists, Thomas Pennant and Daines Barrington. Pennant, like White, lived his life in a circumscribed environment, at his family estate near Whitford in Flintshire. Nonetheless, he wrote acclaimed books on zoology, some of which influenced Samuel Johnson, and maintained a correspondence with many of the leading scientific figures of the day. Daines Barrington was a lawyer, antiquary and naturalist, and in his time he served as the vice president of the Royal Society. White's letters to them both were a process of sharing discoveries and ideas.

As to White himself, the bare facts of his life may be quickly sketched, and point to an outwardly uneventful life. He was born in Selborne on the 18 July 1720, went to school at the nearby Hampshire town of Basingstoke, and was admitted to Oriel College, Oxford in 1739, becoming in turn a fellow of his college and senior proctor of the university. Meanwhile, he had been ordained, and after his return to Selborne in 1755, he held a curacy in the nearby parish of Faringdon until 1784, when he was appointed curate at Selborne, where he lived in a house called The Wakes, (now a museum) until his death on 26 June 1793, four years after the publication of his great book. On 2 January 1769, White wrote a letter from his home in Selborne, Hampshire to Thomas Pennant about the unique sound of the nightjar, a nocturnal bird that was surrounded by considerable superstition. The bird's 'music' was – and is – unique, a strange churring sound which White describes in his letter as being produced through

---

[31] G. White, *The Natural History of Selborne* (London: Ducimus Books, 1974), p. 66.

the windpipe, 'just as cats purr'. The bird's 'music' was – and is – unique, a strange churring sound which White describes in his letter as being produced through the windpipe, 'just as cats purr'.[32] He then adds a paragraph of description that uniquely demonstrates the quality of the nightjar's pitch and timbre:

> You will credit me, I hope, when I assure you, that, as my neighbours were assembled in an hermitage on the side of a steep hill where we drink tea, one of these churn-owls came and settled on the cross of that little straw edifice, and began to chatter, and continued his note for many minutes; and we were all struck with wonder to find that the organs of the little animal, when put in motion, gave a sensible vibration to the whole building![33]

He also notes the punctuality of the start of its 'song', 'just at the report of the Portsmouth evening gun, which we can hear when the weather is still.'[34] Although this almost casual remark is an aside to the main issue in the letter, that of the bird's sound, it is an authentic witness that demonstrates powerfully how sound was transmitted across the land in his day. The recorded memory of a sound that has faded beyond us through time, is preserved for us in the words of White, a moment held that helps us whenever we seek to personalize how the Hampshire village of Selborne sounded in the eighteenth century. Turning to the place itself, even a visitor today will notice that there is a special quality to the soundscape between the great wooded hill known as 'the Hanger', one of the highest points in the county of Hampshire, and White's house, The Wakes. It is as though the hill creates a backdrop, making it like a soundstage that reflects everything in the village and valley below, giving the impression that voices and domestic sounds are acoustically projected onto a screen, and amplified (Perhaps this affected his sound impression of the grasshopper-lark, mentioned earlier?) For all the years and the tourists, this same soundstage holds what seems to be a strong sense of White and his quiet studies, and a timeless whisper of a presence. The village and the book that immortalized it continue to embrace one another, so that it is possible to read White, look up, listen and feel kinship with his impressions. His world was this place, and he wrote thousands of words about it, not only in his *Natural History*, but also in his journals, which will surely delight anyone whose heart has been caught by the more famous book. Here, for example, is his entry for 2 February 1787, six years before his death: 'Storm-cock sings. Brown

---

[32] Ibid., p. 93.
[33] Ibid.
[34] Ibid., p. 93

wood-owls come down from the hanger in the dusk of the evening, and sit hooting all night on my wal-nut [*sic*] trees. Their note is like a fine vox humana, and very tuneable.'[35] The musical notes of birds, and in particular owls, was a subject of fascination to White.

On 1 August 1771, he wrote:

> A friend remarks that many (most) of his owls hoot in B flat; but that one went almost half a note below A. The pipe he tried their notes by was a common half-crown pitch-pipe, such as masters use for tuning of harpsichords; it was the common London pitch.
>
> A neighbour of mine, who is said to have a nice ear, remarks that the owls about this village hoot in three different keys, in G flat or F sharp, in B flat, and A flat. He heard two hooting to each other, one in A flat, and the other in B flat[36]

On 8 July 1773, he returns to the question of owls and sounds:

> White owls seem not (but in this I am not positive) to hoot at all; all that clamorous hooting appears to me to come from the wood kinds. The white owl does indeed snore and hiss in a tremendous manner; and these menaces well answer the intention of intimidating; for I have known a whole village up in arms on such an occasion, imagining the churchyard to be full of goblins and spectres. White owls also often scream horribly as they fly along; from this screaming probably arose the common people's imaginary species of screech-owl, which they superstitiously think attends the windows of dying persons.[37]

White's is the poetry of information, and although his prose is spare, it contains the beauty of the place in which it grew. Indeed, if he tells us that the owls made music like a 'vox humana', we may assume no hyperbole or poetic licence. This makes the sound picture all the truer and more impressive. His biographer, Richard Mabey, has expressed this perfectly:

> In more than fifteen thousand daily entries [in the journal] it is hard to find a single one which can be interpreted as an explicit reference to White's feelings about the natural world. Despite this – though it is really because of this – their imaginative strength and *implicit* emotional content can be very powerful and

---

[35] G. White, 'Journal' in *Gilbert White's Year*, J. Commander (ed.) (Oxford: Oxford University Press, 1982), p. 19.
[36] G. White, *Natural History*, p. 187.
[37] Ibid., p. 198.

clear. [In a phrase or two] he can catch the essence of a moment, or a whole season.[38]

As if to confirm the blend of art and science in his own work, White showed himself to be also a poet, and some editions of *Selborne* contain a smattering of his verses, such as 'Invitation to Selborne', 'Selborne Hanger', 'On the Rainbow' and 'A Harvest Scene.' Here, as in the body of his famous prose, there are moments of sound recording, for example in his poem 'On the Dark, Still, Dry, Warm Weather', ('Occasionally happening in the winter months'):

> While high in air, and poised upon his wings,
> Unseen, the soft, enamour'd woodlark runs
> Through all his maze of melody; the brake,
> Loud with the blackbird's bolder note, resounds.
> Soothed by the genial warmth, the cawing rook
> Anticipates the spring, selects her mate,
> Haunts her tall nest-trees, and with sedulous care
> Repairs her wicker eyrie, tempest-torn.[39]

The dead leave us their art, their words and their ideas, but until recent history, they have not left us their physical voices, and the places they occupied or frequented have in many cases been so changed, and in some cases, destroyed, that in order to 'hear' them and their world, we must become imaginative receivers, as Bruce R. Smith has so well put it, to '"un-air" sounds that have faded into the air's atmosphere'.[40]

Perhaps a rural heaven-on-earth *did* exist, but then so did its counterpart, and by the time White was writing his book, this idea of country life was already under assault from the forces of progress, reflected in the reality of class and poverty. Mechanization would soon impose itself in unimagined ways; already the dull roaring of towns and cities was bleeding into field and hedgerow, continuing the change of the sound of the world, while affecting our ability – our increasing *in*ability – to hear the detail of the small sounds of nature, and the events that had punctuated stillness in Gilbert White's day and beyond.

On Saturday, 4 May 1785, White wrote in his journal: 'Flisky clouds about … Crickets sing much on the hearth this evening: they feel the influence of moist air,

---

[38] R. Mabey, *Gilbert White's Year*, p. 8.
[39] Ibid., pp. 7–8.
[40] Smith, in M. Bull and L. Black, *The Auditory Culture Reader* (New York: Berg, 2003), p. 129.

and sing against rain.'[41] He was a man who observed and drew informed deductions from what he saw and heard. In that same year, the poet, William Cowper published his masterpiece, *The Task*, a poem in six books, written in blank verse. At the start of the last book, entitled 'The Winter Walk at Noon', there is a passage that proves him to be a man who, like White, made connection. Here, in eighteen lines, he finds a crucial link between sound and memory, something we must take into the next chapter. His metaphorical 'microphone' records the sounds of the church bells of his home town of Olney in Buckinghamshire, but it is in his heart and mind that the meaning is lodged:

> There is in souls a sympathy with sounds,
> And as the mind is pitch'd the ear is pleased
> With melting airs or martial, brisk or grave.
> Some chord in unison with what we hear
> Is touch'd within us, and the heart replies.
> How soft the music of those village bells
> Falling at intervals upon the ear
> In cadence sweet! now dying all away,
> Now peeling loud again and louder still,
> Clear and sonorous as the gale comes on.
> With easy force it opens all the cells
> Where memory slept. Wherever I have heard
> A kindred melody, the scene recurs,
> And with it all its pleasure and its pains.
> Such comprehensive views the spirit takes,
> That in a few short moments I retrace
> (As in a map the voyager his course)
> The windings of my way through many years.[42]

In this beautiful passage, Cowper touches on the continuing power of sound to evoke memory, a factor that must be central to any philosophical exploration of auditory aesthetics. Further, *The Task* is a key to the next generation of writers; the conversational diction, as well as the thinking behind it, was a major influence on Samuel Taylor Coleridge and William Wordsworth, notably in the latter's 'Tintern Abbey' and 'The Prelude'. Above all, Cowper teaches us how to listen rather than simply hear; his church bells rang across the parish, its lanes and fields, and they ring on in memory through his words.

---

[41] G. White (ed.), F. Greenoak, and R. Mabey, *The Journals of Gilbert White, Volume Three: 1784–1793* (London: Century, 1989), p. 88.
[42] W. Cowper, *The Poetical Works of William Cowper* (London: Frederick Warne, 1893), p. 310.

7

The romantic poetry of listening

The 'long' eighteenth century saw a complex relationship between the archetypal bucolic sounds of the apparently perpetual natural world, and the invasion of that world by the noise of industry. Today, we are never far from the glow in the night sky formed by a city beyond the horizon, the hum of a motorway or the drone of aircraft. To gain a sense of the profound silence of nightscapes in the countryside of the pre-industrial past is to imaginatively project an idea onto a landscape sonically affected by modern sounds. Furthermore, we have come to view and to hear the incursions of industry and urbanization on an idealized concept of the countryside, in the light of the subsequent environmental crisis. Yet, it is surprising to realize that there was sometimes an ambivalence in attitudes to what many would consider desecration, and for some even an exhilaration in this growing dynamism. Sheffield's Ebenezer Elliott (1781–1849), for example, known as 'the corn-law rhymer', could celebrate the heroic spectacle of the new industry:

> Oh, there is glorious harmony in this
> Tempestuous music of the giant, Steam,
> Commingling growl, and roar, and stamp, and hiss,
> With flame and darkness! Like a cyclops' dream,
> It stuns our wandering souls, that start and scream
> With joy and terror
> It is beneficent thunder[1]

On the other hand, industry's invasion of rural landscapes was more often the subject of outrage rather than approbation, a view that was to increase as the nineteenth century progressed. As early as 1787, Anna Seward (1742–1809), the daughter of the vicar of the Derbyshire village of Eyam, viewed the industry of iron workings in the Shropshire valley of Coalbrookdale in terms

---

[1] E. Elliott, 'Steam at Sheffield', quoted in M. Drabble, *A Writer's Britain* (London: Thames and Hudson, 2009), p. 201.

of what decades of factory development had done to the pastoral perfection of the valley in which Abraham Darby ultimately smelted iron ore for the first time. This process used coal drawn from the side of the valley, thus making the gorge on the Severn, already spectacular, one of the most important places in the Industrial Revolution. For Seward however, as for many others, there was a noisy conflict between the promise of the future, and a nostalgia for a gentler, more lyrical past:

> Black sulphurous smoke, that spread their veils
> Like funeral crape [sic] upon the sylvan robe
> Of thy romantic rocks, pollute thy gales,
> And stain thy glassy floods; – while o'er the globe
> To spread thy stores metallic, this rude yell
> Drowns the woodland song, and breaks the Poet's spell.[2]

Meanwhile, in certain areas of writing, there was still a nostalgia for the distant pastoral past, as suggested in Seward's poem. It is not the purpose of this book to provide an in-depth critical survey of the evolution of rural writing, apart from where we may explore the growth and reflection of sonic awareness in relation to the natural world. Nor is this book a social history, examining in detail the relationship between classes and the effects of industry on the changing landscape. Sound is and remains our theme here, as 'recorded' and 'heard' through written text. That said, both cultural and social context must of necessity enter into our consideration as they become either directly or tangentially relevant to the subject. As we have discussed, time and cultural shifts make for a complex mix when defining the auditory backdrop, and foreground keynote sound motifs of any particular era. Awareness of change as reflected in writing is always layered, and nostalgia does not belong exclusively to a particular time, while in a sense, it is part of a certain type of mindset inherent in *every* era. We have already noted it as a recurrent theme, present in some seventeenth- and eighteenth-century rural writing, just as it would be in the years surrounding the world wars of the twentieth century. Although the Eden myth as a metaphor is a convenient shorthand within a Christian context, a yearning for the simpler pastoral fiction of an idealized rural past – a 'Golden Age' – was already present in Hesiod (*c.* 750–650 BC), Theocritus (*c.* 300–*c.* post-260 BC) and Virgil (*c.* 70–19 BC).

---

[2] A. Seward, 'Sonnet: To Colebrooke Dale' in R. Lonsdale (ed.), *Eighteenth Century Women Poets* (Oxford: Oxford University Press, 1989), p. 316.

Raymond Williams has said that, even for Hesiod, 'at the beginning of country literature, it is already far in the past' although 'it is in the Hellenistic world of the third century before Christ that "pastoral", in any strict sense, emerges as a literary form'.³ And when it does, with it comes a sense of yearning for the lost whispering groves of Arcadia, as in the Idylls of Theocritus:

> All rich delight and luxury was there:
> Larks and bright finches singing in the air;
> The brown bees flying round about the well;
> The ring-dove moaning⁴

It is this pastoral landscape, transferred to the sculpted landscaped English gardens of Charles Bridgeman and his successors⁵, in which Anne Finch's two Arcadian shepherdesses languidly discussed love and young swains in the 1799 poem quoted in the previous chapter. Nostalgia in rural writing has continued to be a contentious issue as we have seen, and the idea of landscape as a sort of manicured 'park' is often too easily translated into writing. White could have written about Arcadia, indeed, there were pastoral frolics in and around Selborne of which he was a part in his youth, but over and above this his book and the researches of which it is the most famous product mark him as the first ecologist and father of nature writing. He was also ordained as a curate, although, as Mabey has said, 'his Christian vocation is barely visible beneath his Enlightenment rationality and early Romantic leanings'.⁶

Scientific study and progress did not of necessity have to be noisy, and could still find a complement in poetic writing. Also as Williams reminds us, it is appropriate to establish a correlation between Gilbert White on one hand, and the poets William Wordsworth, Samuel Taylor Coleridge and John Clare on the other, in their 'intense devotion to watching and describing nature'. Yet, in Williams's words, White's 'close observation and description is of a separated object, another creature', while in Wordsworth et al., that separation 'is mediated by a projection of personal feeling into a subjectively particularized and objectively generalized Nature'.⁷ In the meantime, there was an ever-growing

---

³ R. Williams, *The Country and the City* (London: Vintage, 2016), p. 20.
⁴ Theocritus, quoted in R. Williams, p. 22.
⁵ Charles Bridgeman (1690–1738) is seen as the founder of landscape gardening later developed by the likes of Lancelot 'Capability' Brown and Humphry Repton.
⁶ R. Mabey, 'Some Key Stations in Gilbert White's Life' in *Turning the Boat For Home* (London: Chatto and Windus, 2019), p. 35.
⁷ R. Williams, *The Country and the City* (London: Vintage, 2016), p. 193.

distinction and tension between urban and rural sounds. There were writers for whom the countryside was the sonic common denominator, the reference point upon which and by which other sounds were compared and contrasted. On the other hand, there were those who came to the country either as a short-term recreational interlude, or for a longer sojourn, as a kind of creative retreat.

Leigh Hunt (1784–1857), as cofounder of *The Examiner* and the centre of the Hampstead-based group that included William Hazlitt and Charles Lamb, often wrote in his essays of the trials of urban living in terms of clean air and noise pollution. Given his work place, he was also well equipped to draw effective comparisons with the quieter environments that then lay within relatively easy reach of London. In his essay, 'Fine Days in January and February' he offers a vivid aural juxtaposition. First, polite city life:

> If you can hear anything but noise, you hear sparrows ... Then in the noisier streets, what a multitude and a new life! What horseback! What promenading! What shopping and giving good day! Bonnets encounter bonnets ... The yellow carriages flash in the sunshine; lapdogs frisk under their asthmas.
>
> Then in the country, how emerald the green, how open-looking the prospect! ... Voices of winter birds are taken for new ones; and in February the new ones come – the thrush, the chaffinch, and the wood-lark. Then rooks begin to pair; and the wagtail dances in the lane. As I write this article, the sun is on our paper, and chanticleer[8] ... seems to crow in a very different style, lord of the ascendant.[9]

What becomes clear is that not only is the air clearer in the country, but natural sounds, while proliferating, are much more distinguishable. The sound layering of the world is becoming more critical as 'civilization' makes its noisy presence felt, and our ability both to hear and interpret the rural soundscape produces a vivid response. Leigh Hunt's 'microphone' can only capture the prevailing sounds.

Meanwhile, in the West Country and the Lake District, other mental and psychic 'recording equipment' had been tuning with ever-more sensitivity to sounds and their implications for the inner being in relation to the natural

---

[8] Chanticleer: Name given to a domestic cock, as heard in farmyards and so on, especially in fairy tales, as the vocalization of morning.
[9] L. Hunt, 'Fine Days in January and February' in A. Symons (ed.) *The Essays of Leigh Hunt* (London: J. M. Dent, 1903), pp. 136–7.

world. William Wordsworth (1770–1850) and Samuel Taylor Coleridge (1772–1834), although born earlier, were both contemporaries of Hunt; indeed Coleridge died in Highgate, adjacent to Hunt's Hampstead. In their writings, it was often sounds heard from elsewhere that informed the visible landscape, rather than the immediate and obvious primary sources around them. During the Wordsworths' Devon period, in an entry in her *Alfoxton*[10] *journal*, dated 23 January 1798, William's sister Dorothy (1771–1855) wrote: 'The sound of the sea distinctly heard on the tops of the hills, which we could never hear in summer. We attribute this partly to the bareness of the trees, but chiefly to the absence of the singing of birds, the hum of insects, that noiseless noise which lives in the summer air.'[11] The year 1798 was a key year for Wordsworth and Coleridge, in that it saw the first publication of their great joint poetic project, *Lyrical Ballads*. Coleridge contributed fewer poems to the collection than did Wordsworth, but his poem 'The Nightingale' makes a strong point against the perceived *idea* of birdsong, and for the reality of it. The accepted literary conceit for the song of the nightingale was that it was melancholy, but, Coleridge argues in the poem, 'In Nature there is nothing melancholy.' Nevertheless the literary world, he goes on, is full of poets who perpetuate the cliché without actually listening to the real thing, and allowing it to work on personal ideas and feelings. As he says:

> Many a poet echoes the conceit;
> Poet who hath been building up the rhyme
> When he had better far have stretched his limbs
> Beside a brook in mossy forest-dell,
> By sun or moon-light, to the influxes
> Of shapes and sounds and shifting elements
> Surrendering his whole spirit, of his song
> And of his fame forgetful! so his fame
> Should share in Nature's immortality & c.[12]

---

[10] Alfoxton, variously known as Alfoxton House and Alfoxton Park: An eighteenth-century house in Holford, Somerset. William and Dorothy Wordsworth lived here from July 1797 to June 1798, during their friendship with Samuel Taylor Coleridge, shortly before their move to the Lake District.

[11] D. Wordsworth, 'Alfoxton Journal' in T. Maybery (ed.) *Coleridge and Wordsworth in the West Country* (Stroud: Alan Sutton Publishing, 1992), p. 158.

[12] S. T. Coleridge, 'The Nightingale: A Conversational Poem' in H. W. Garrod (ed.) *Coleridge: Poetry and Prose* (Oxford: Clarendon Press, 1954), pp. 113–14.

Of this poem, Jeremy Hooker has commented: 'It is the melancholy man who apprehends the nightingale's song as melancholy, thereby revealing nothing of the truth about nightingales or nature, but displaying the human propensity for egotism.'[13] (John Clare, who we shall consider in the next chapter, would have heartily agreed.) Hooker goes on to develop the point, reflecting on the importance of the line, ' "And many a poet echoes the conceit". The effect of this statement is to distinguish the new poetry of *Lyrical Ballads* from imitative verse. The former arises from self-surrender, in which the poet shares in the spirit of nature.'[14] Artificial versifiers do Nature and mankind a profound wrong in perpetuating a false message, because they are the source of information for 'youths and maidens most poetical' who, while spending their time in 'ball-rooms and hot theatres' know no better, and therefore 'heave their sighs/O'er Philomena's pity-pleading strains.' Then, he turns to William and Dorothy, as kindred spirits:

> My friend, and thou, our Sister! we have learnt
> A different lore: we may not thus profane
> Nature's sweet voices, always full of love
> And joyance! 'Tis the merry Nightingale
> That crowds, and hurries, and precipitates
> With fast thick warble his delicious notes,
> As he were fearful that an April night
> Would be too short for him to utter forth
> His love-chant, and disburthen his full soul
> Of all its music! & c.[15]

The poem is a lesson in location recording; do not accept second-hand information, he exhorts, but gather the reality of the experience in the field, in order to gain a true interpretation. More, beware of stereotypes and archetypes, especially when considering the natural world. There are easy, lazy shortcuts when it comes to playing on shallow emotions, but in the end, we listen with our mind, and preconceptions can only cloud experience. Coleridge feels he understands what makes birds sing, and in his 1802 poem, 'Answer to a Child's Question', he takes on the matter head-on in his answer:

---

[13] J. Hooker, ' "Awakening the Mind's Attention": Lyrical Ballads and the Art of Seeing' in *Art of Seeing: Essays on Poetry, Landscape Painting and Photography* (Swindon, Shearsman Books, 2020), p. 25.
[14] Coleridge, 'The Nightingale.'
[15] Ibid.

> Do you ask what the birds say? The sparrow, the dove,
> The linnet and thrush say, 'I love and I love!'
> In the winter they're silent – the wind is so strong;
> What it says, I don't know, but it sings a loud song.
> But green leaves, and blossoms, and sunny warm weather,
> And singing, and loving – all come back together. & c.[16]

Frequently, Coleridge and Wordsworth linger on the sound of owls. Wordsworth, in his poem, 'The Idiot Boy' seems, through the mind of the boy himself, almost at times to give the owl the responsibility of being the auditory manifestation of the silent moon:

> The moon is up – the sky is blue,
> The owlet in the moonlight air,
> He shouts from nobody knows where;
> He lengthens out his lonely shout,
> Halloo! halloo! a long halloo![17]

William Wordsworth's contemplation of sound was profound, and his attempts at replicating the tones, pitches and vibrations of the environment drove him to use every tool in the poet's box with the utmost skill. The move to Grasmere in the Lake District opened up a world of echoes, both sonic and mnemonic, to which he and his sister became highly tuned. In an entry in Dorothy's journal for example, written during a spell of very hot weather in July 1800, there is this:

> After tea we rowed down to Loughrigg Fell, visited the white foxglove, gathered wild strawberries, and walked up to view Rydale. We lay a long time looking at the lake, the shores all embrowned with the scorching sun ... The lake was now most still ... We heard a strange sound in the Bainriggs wood as we were floating on the water. It *seemed* in the wood, but it must have been above it, for presently we saw a raven very high above us – it called out and the dome of the sky seemed to echo [sic] the sound – it called again and again, and the mountains gave back the sound, seeming as if from their centre a musical bell-like answering to the bird's harsh voice. We heard the call of the bird and the echoe after we could see him no more.[18]

---

[16] S. T. Coleridge, 'Answer to a Child's Question' in *The Works of Samuel Taylor Coleridge* (Ware: Wordsworth Editions, 1994), p. 386.
[17] W. Wordsworth, 'The Idiot Boy' from *Lyrical Ballads* (London: Penguin Books, 1999), p. 81.
[18] D. Wordsworth, *Grasmere Journal* (Oxford: Oxford University Press, 2008), p. 14.

The combination of the still water and the sound-containing landscape of reflecting mountainsides is a potent one; in certain climatic conditions, everything becomes amplified. Looking over Rydale in October, 1800, Dorothy noted that 'the lowing of the cattle was echoed by a hollow voice in Knab Scar'.[19] This sonic phenomenon often found its way into Wordsworth's poetry of the Lake District, as in the following famous incident from his childhood, in which the boy steals a boat to go rowing out onto to lake. The lowering mountains seem to threaten and reprimand him as he rows out, but the sounds of his actions across the still water and air reflect like conscience:

> It was an act of stealth
> And troubled pleasure, nor without the voice
> Of mountain-echoes did my boat move on[20]

'All shod with steel,/ We hissed along the polished ice' he writes in the 'Skating' episode, quoted previously, and the sound of the skates in the still, cold air is palpable. The onomatopoeia offers us the ability to listen *through* the words to the *actual* sound in its 'live' moment, as near to a microphone as a phrase could ever get:

> With the din
> Smitten, the precipices rang aloud;
> The leafless trees and every icy crag
> Tinkled like iron; while far distant hills
> Into the tumult sent an alien sound
> Of melancholy not unnoticed, while the stars
> Eastward were sparkling clear[21]

Dorothy's role in William's work has to be acknowledged. It was she who often pointed things out, or reminded him, or, through her journals, preserved a moment he would later explore in a poem: 'She gave me eyes, she gave me ears', he wrote in tribute.[22] For the Wordsworths, their immediate vicinity became a sound studio, with every aural faculty heightened, and informed by the spectacular visuals with which the scenery surrounded them. Above all, it was the stillness containing each tiny sound as it fell and reverberated that caught

---

[19] Ibid., p. 28.
[20] W. Wordsworth, *The Prelude* (Oxford: Clarendon Press, 1959), p. 25.
[21] Ibid., pp. 27–9.
[22] W. Wordsworth. In D. Wordsworth, *Grasmere Journal* (Quoted by P. Woolf) in Introduction (Oxford: Oxford University Press, 2008), p. xvi.

their attention, and they became increasingly susceptible as their listening heightened. In late April 1802, on a beautiful morning, Dorothy and William lay in the garden, listening, eyes closed, 'to the waterfalls and the birds' as Dorothy recorded in her journal:

> There was one waterfall after another – it was the sound of waters in the air – the voice of the air. William heard me breathing and rustling now and then but we both lay still, and unseen by one another – he thought it would be sweet thus to lie so in the grave, to hear the peaceful sounds of the earth and just to know that ones' dear friends were near.[23]

In June that year, after a sleepless night (the Wordsworths were eccentric in their time keeping when it came to the boundaries between day and night), Dorothy noted the minutiae of the emerging morning sounds:

> The shutters were closed, but I heard the birds singing. There was our own thrush shouting with an impatient shout – so it sounded to me. The morning was still, the twittering of the little birds was very gloomy. The owls had hooted a quarter of an hour before. Now the cock was crowing. It was near daylight. I put out my candle and went to bed.[24]

Coleridge also records a conversation between owl and cock at the start of the 1797 gothic tale of 'Christabel':

> 'Tis the middle of the night by the castle clock,
> And the owls have awakened the crowing cock;
> Tu–whit! – Tu–whoo!
> And hark, again! the crowing cock,
> How drowsily it crew.[25]

In his poem, 'On the Power of Sound' Wordsworth speaks of the ear an 'organ of vision'; it is a hymn to the aural sense. In this context the natural world, memory and the human imagination, are part of the transmission equipment, but the ear is more than a receiver here. In his argument preceding the poem, which was published in 1835, but written seven years earlier, Wordsworth considers the human ear 'as occupied by a spiritual functionary, in communion with sounds, individual or combined in studied harmony'. This is a system of 'call and response', an active and dynamic relationship which amounts to the

---

[23] D. Wordsworth, *Grasmere Journal*, p. 92.
[24] Ibid., p. 112.
[25] S. T. Coleridge, 'Christabel' in *Coleridge: Poetry and Prose*, p. 79.

sort of listening advocated over two centuries later by sound artists such as the composer and musician, Pauline Oliveros. The poem is included in his collection, *Poems of the Imagination*, in which are also featured verses on the cuckoo, the nightingale, the sky lark and water fowl. In this fourteen-stanza poem, however, he pulled these together into a soundscape that included human and non-human contributors, from within and beyond the horizon, such as a desert lion, Syrian landscapes, the murmurings in a convent and a prayer for salvation breathed on a stormy sea. All this and more feeds into the ear; 'Thy functions are ethereal', the poem begins, 'As if within thee dwelt a glancing mind' Wordsworth links physical and imaginative listening through the ear, which itself seems to contain a driving spiritual intelligence:

> The headlong streams and fountains
> Serve Thee, invisible Spirit, with untired powers;
> Cheering the wakeful tent on Syrian mountains,
> They lull perchance ten thousand thousand flowers.
> *That* roar, the prowling lion's *Here I am*,
> How fearful to the desert wide!
> That bleat, how tender! of the dam
> Calling a straggler to her side.
> Shout cuckoo! – let the vernal soul
> Go with thee to the frozen zone;
> Toll from thy loftiest perch, lone bell-bird, toll!
> At the still hour to Mercy dear,
> Mercy from her twilight throne
> Listening to nun's faint throb of holy fear,
> To sailor's prayer breathed from a darkening sea,
> Or widow's cottage-lullaby.[26]

Imaginative sound needs no passport, crosses every geographical boundary. This pivotal poem, and the thinking that led to it, as articulated elsewhere in the Wordsworths' writings, provides a staging post, a landmark in the textual communication of sound, its physical reception by the human ear and its interpretation by the mind. Both Wordsworth and Coleridge expressed vividly at various points in their work, the idea that the physical world is a manifestation of something else, and what better means to show that than through the

---

[26] W. Wordsworth, 'On the Power of Sound,' (Stanza ll) in *The Poetical Works of William Wordsworth* (London: Henry Frowde/Oxford University Press, 1909), pp. 232–3.

invisible medium of sound? In 1799, Coleridge had written of an experience in the Harz Forest[27], descending from 'Brocken's sovran height' through dense fir woods where

> But seldom heard,
> The sweet bird's song became a hollow sound:
> And the breeze, murmuring indivisibly,
> Preserved its solemn murmur most distinct
> From many a note of many a waterfall,
> And the brook's chatter[28]

The forest works its spell, and Coleridge, as he listens on his walk down into the valley, senses through these sounds, 'that outward forms, the loftiest, still receive/The finer influence from the life within;– /Fair cyphers else'.

We must also consider, when 'hearing' the sounds through these poems, the voices – accent, dialect, tones and quality of speech – with which the writers themselves spoke. Just as Shakespeare wrote in the voice of his time, and as we discussed, sometimes possibly the *actual* voices of his players in mind, with puns and wordplay reflected in language sounds from which time has distanced us, so we should take into account the human sounds with which poets such as Wordsworth and Coleridge 'heard' the language they wrote, and how they articulated them in their readings and conversations. We might assume that Wordsworth possessed a northern English accent, while William Hazlitt remembered of Coleridge that 'his voice rolled on the ear like the pealing organ, and its sound alone was the music of thought ... Shall I, who heard him then, listen to him now? Not I! ... that spell is broken; but still the recollection comes rushing by with thoughts of long-past years, and rings in my ears with never-dying sound.'[29] Likewise, anyone who had known the poet in life, might have had a subsequent 'silent' reading of his work coloured by the memory of his voice. Subsequent generations of course, do not have the benefit of Hazlitt's personal experience, but the attentive reader may detect it in the syntax and sound/word usage of a poem by a long-dead author. There are clues as to what the writer was 'hearing' in his or her head, both of the actual sonic world they were seeking to replicate, and the inner ear of language-sound of both themselves and of the

---

[27] The Harz (sic) or Hartz is a highland area in northern Germany.
[28] S. T. Coleridge, 'Lines' in *Coleridge: Poetry and Prose*, pp. 121–2.
[29] W. Hazlitt, Essay: 'Poetry in General', quoted by A. Piette. *Remembering and the Sound of Words* (Oxford: Oxford University Press, 2004), p. 35.

environment in which they lived and worked. This would have been relevant to the thinking of Coleridge and Wordsworth; in the 'Advertisement' to the *Lyrical Ballads*, written for the 1798 edition, Wordsworth made the point that 'the majority of the following poems are to be considered as experiments. They were written chiefly with a view to ascertain how far the language of conversation in the middle and lower classes of society is adapted to the purpose of poetic pleasure.'[30] The vocal characteristics of the poems in *Lyrical Ballads* become evident further into 'Advertisement', when Wordsworth refers to the poem called 'The Thorn', which, 'as the reader will soon discover, is not supposed to be spoken in the author's own person: the character of the loquacious narrator will sufficiently show itself in the course of the story'.[31] It is a remarkable poem, and clearly fulfils the brief set as part of the 'experiment' behind the *Ballads*. Indeed, there are pre-echoes of the poems of Edward Thomas and Robert Frost in the conversational tone of the writing:

> There is a thorn: it looks so old,
> In truth you'd find it hard to say,
> How it could ever have been young,
> It looks so old and grey.
> No higher than a two-years' child,
> It stands erect this aged thorn;
> No leaves it has, no thorny points;
> It is a mass of knotted joints,
> A wretched thing forlorn.
> It stands erect, and like a stone
> With lichens it is overgrown. (& c.)[32]

There is no doubt about a 'heard' voice here from the start; 'The Thorn' reads almost like a transcription of spoken language. Yet it is full of art. Wordsworth wrote a long and rather wordy 'Preface', often not included in the modern editions of the collection. It contains the famous passage, articulating his view 'that poetry is the overflow of powerful feelings: it takes its origin from emotion recollected in tranquillity: the emotion is contemplated till by a species of reaction the tranquillity gradually disappears, and an emotion, similar to that which was before the subject of contemplation, is gradually produced, and does

---

[30] W. Wordsworth, 'Advertisement', *Lyrical Ballads*, p. v.
[31] Ibid., p. vi.
[32] W. Wordsworth, 'The Thorn' from *Lyrical Ballads*, p. 63.

itself *actually exist in the mind*' [My italics].³³ This idea, that the experience and feeling can be conjured back, so that the event is not only recollected but also *actually* exists again, is a potent one, particularly when applied to sound; it is 'recording' in every sense of the word.

The range of styles within the book is considerable; the first edition opens with Coleridge's famous 'Rhyme of the Ancient Mariner' (a fact that Wordsworth later regretted, feeling its length and style were out of keeping with the rest of the collection). Running through the poems overall, however, is a lyric voice that drives the idea of a oneness between the alert senses and the physical world, sometimes through storytelling, sometimes through meditative attentiveness. Wordsworth's ear as an 'organ of vision' is tuned to the signals reaching him from his surroundings, and at the climax of his great poem, 'Lines Written a few miles above Tintern Abbey', conceived in July 1798 during a visit to the Wye Valley after absence, there comes a key passage in which the poet acknowledges that he is right to trust his sensory responses as a kind of portal through which the spirit of Place and his own being pass, meet and commingle:

> Therefore am I still
> A lover of the meadows and the woods,
> And mountains; and of all that we behold
> From this green earth; of all the mighty world
> Of eye and ear, both what they half-create,
> And what perceive; well pleased to recognize
> In nature and the language of the sense,
> The anchor of my purest thoughts, the nurse,
> The guide, the guardian of my heart, and soul
> Of all my moral being.³⁴

Here again there is this powerful idea that the eye and ear are both receivers and imaginative creators, the key to the relationship between 'nature and the language of the sense'. It was a major legacy.

In April 1818, John Keats wrote what his biographer Andrew Motion has called 'a great prose hymn to progress'³⁵ in a long letter to his friend John Hamilton Reynolds, during which he discusses his ideas of growing consciousness within a human life. It is a kind of 'seven ages of man' reduced

---

[33] W. Wordsworth, *Prefaces to the Lyrical Ballads* (London: Thomas Nelson, 1937), p. 38.
[34] W. Wordsworth, *Lyrical Ballads*, p. 112.
[35] A. Motion, *Keats* (London: Faber and Faber, 1997), p. 254.

to key staging points in thinking. He speaks of thought becoming gradually darkened, 'and at the same time on all sides many doors are set open ... all leading to dark passages'. He goes on: 'We are in a mist ... We feel the burden of the Mystery. To this point was Wordsworth come, as far as I can conceive when he wrote "Tintern Abbey" and it seems to me that his genius is explorative of those dark passages. Now, if we live, and go on thinking, we too shall explore them'.[36] Keats was twenty-three years old at this time, with less than three years left, and when 'Tintern Abbey' was written he had been a young child. In 1818, Wordsworth represented a previous generation, whose thinking was to be by turns venerated, questioned and built on. The ideas expressed in this letter are useful when considering what happened next in his creative life. Keats at this time was already suffering from persistent sore throats, the beginning of his last illness, yet was still to write some of his greatest poems expressing sound consciousness, notably his 'Ode to a Nightingale', the 'Ode on a Grecian Urn' and the 'Ode to Autumn', all of which were created in his miraculous year of 1819. In these poems there are some 'dark passages' [perhaps we may interpret 'dark' in one sense as also 'still' or 'silent'?] from which come sounds both of nature and of imaginative listening. We know that Wordsworth admired Keats's 'Ode to a Nightingale', and it may be that the 'dark passages' down which the poem peers and listens were of influence on his own thinking, as well as feeding it in later poems such as 'On the Power of Sound'. There is the imaginative journey which the song of Keats's bird initiates, growing trance-like; darkness removes the power of outward sight and sound, and something else takes over. 'Darkling I listen' says Keats in the next stanza, and he thinks of his own mortality, yet by stanza seven, the song of the nightingale becomes a time warp, a voice across the aeons, linking his Hampstead garden with the past and its attic literature celebrating this song, epitomized by and personified by this very bird:

> The voice I hear this passing night was heard
> In ancient days by emperor and clown:
> Perhaps the self-same song that found a path
> Through the sad heart of Ruth, when, sick for home,
> She stood in tears amid the alien corn;[37]

[36] J. Keats, letter to John Reynolds, quoted in *Motion*, p. 255.
[37] 'Ruth'. This refers to the story, told in the eponymous Biblical book, of a Moabite woman who married an Israelite. After the death of all the male members of her family (her husband, father-in-law and brother-in-law), she stays with her mother-in-law, Naomi, and moves to Judah with her. Ruth is seen as a symbol of abiding loyalty and devotion.

> The same that oft-times hath
> Charm'd magic casements, opening on the foam
> Of Perilous seas, in faery lands forlorn.[38]

From which point the final stanza opens with one of the greatest sound connections in poetry:

> Forlorn! The very word is like a bell
> To toll me back from thee to my sole self![39]

Keats is concerned with recording sound in text: both the hearing, and the feeling it engenders, the associative power of a sound to link with a new thought. In his 'Ode on a Grecian Urn', he takes this further into the imagined sound emitting from a silent object, writing a poem which is as full of sound as any that came from his pen. The key feature within this psychic sound world is the juxtaposition of silence and clamorous sound, a meditation on the relief images of a Greek vase as Keats brushes his imaginative fingers over the braille of the embossed forms depicted and turns up the volume; yet it is all in the fancy, conjured by the brain from dumb witness. This, for the poet, makes the sounds all the clearer and eloquent:

> Heard melodies are sweet, but those unheard
>   Are sweeter; therefore, ye soft pipes, play on;
> Not to the sensual ear, but, more endear'd,
>   Pipe to the spirit ditties of no tone[40]

Keats is *reading* the vase, as he might read a book, but he is also 'playing' it, as if it were a record, a tape or a sound file. It is as though there is in him a silence waiting to be filled, like a reservoir that the sound world flows into as we engage with the sonic circumstances of where we are. In the case of the imaginative sound that is *self*-created (or rather created through interaction between the mind and an otherwise silent source – as with Keats's vase) we approach a deeply personalized area of our being, in which each of us may 'hear' something differently, or at least a variant of the same sound, 'tuned' by our own imagination. These are 'dark passages' that we alone can explore, sensory and imaginative responses on the edge of awareness and understanding, and there is a strand of poetic exploration in Coleridge, Wordsworth, and Keats that toys hauntingly with

---

[38] Ibid., pp. 208–9.
[39] Ibid., p. 209.
[40] Ibid., p. 209. 'Ode on a Grecian Urn', stanza ii.

the idea of the almost seen and almost heard. The natural world plays into this concept, with perspectives of distance, layered sound and ambivalent signals from sonic events occurring within the same environment. (Is that the rushing of water from a nearby stream, or the wind moving the leaves of the trees?) There is sound coming from all directions, and frequently the source is either unseen or unknown – or both. Just as Keats's vase plays out its sound silently, so conversely the natural world often emits sound invisibly. 'The owlet shouts from nobody knows where' writes Wordsworth in 'The Idiot Boy', quoted above, and in the nightingale ode, Keats senses more than observes his surroundings: 'I cannot see what flowers are at my feet,/ Nor what soft incense hangs upon the boughs,/ But, in embalmed darkness, guess each sweet'[41] & c. At the end, the bird's song fades 'Past near meadows, over the still stream, / Up the hill-side; and now 'tis buried deep/In the next valley-glades'[42] & c. There are layers upon layers of sound, and the birdsong moves from a physical event gradually into the imagination. It is like the decaying note of a bell, when the sound moves beyond the sensory capacity to hear it, but not the mind's capability of conjuring it. Finally, below the layers, there is a silence, and within even that, a further profound stillness. At the end of 'On the Power of Sound' Wordsworth asks: 'O Silence! are Man's noisy years/ No more than moments of thy life?'[43]

Keats, in the last stanza of 'To Autumn' tunes his listening – and ours – step by step through variations of sensitivity until the noise is filtered out, leaving in its place a wash of small sounds, loaded with significance. 'Where are the songs of Spring? Ay, where are they? Think not of them, thou hast thy music too.'[44] The world is transformed from that of gaudy bright emerging flowers and the songs of mating birds, to a russet pastoral place of subtle murmurings, evoked in the trance-like last lines of the poem, where the eyes close and the fading year mellows into the pure gentle soundscape of farewell:

> Then in a wailful choir the small gnats mourn
> Among the river sallows, borne aloft
> Or sinking as the light wind lives or dies;
> And full-grown lambs loud bleat from hilly bourn;
> Hedge-crickets sing; and now with treble soft

---

[41] J. Keats, 'Ode to a Nightingale', stanza v.
[42] Ibid., stanza viii.
[43] W. Wordsworth, 'On the Power of Sound', stanza xiv.
[44] J. Keats, 'Ode to Autumn', stanza iii in *Keats: Poetical Works*, p. 219.

> The red-breast whistles from a garden-croft;
> And gathering swallows twitter in the skies.[45]

This is intense, concentrated writing, a field recording that preserves place and time – St Cross water meadows, near Winchester, 1819 – against change; (today, among those changes there is the roar of the M3 motorway as it cuts through the nearby hillside of Twyford Down on its way from Southampton to London). Keats's internal microphone documents for us the way things were, a specimen of a place and time, a mnemonic to transport the mind and a meditation-evoking deep and subtle intimations of mortality. Everything in that final stanza of 'Autumn' evokes our response through sound, partly because it triggers our memory and sense of those sounds gleaned from our own experience, and partly by the cumulative power of the peeling back, through sound under sound, towards a stillness beneath. We do not need to see these things other than with the inner eye; indeed, we do not even need to hear them because the poem itself creates them again, so that the place, the time and the event '*does itself actually exist in the mind*' as Wordsworth suggested. A year after Keats's trip to Winchester and 'Autumn's poetic genesis, his friend Percy Bysshe Shelley (1792–1822) was writing one of his most famous poems in Pisa, 'To a Skylark', 'That from Heaven, or near it,/ Pourest thy full heart'.[46] Sound here is yet again the invisible medium, the ghostly presence that touches us in passing, then moves on through time, fading back into silence down its 'dark passage'. Shelley tries to follow the bird, as sound and light blend and dissolve:

> Thou art unseen, but yet I hear thy shrill delight,
> Keen as are the arrows
>   Of that silver sphere,
> Whose intense lamp narrows
>   In the white dawn clear
> Until we hardly see – we feel that it is there.[47]

The physical, perceived world can be defined through the invisibility of sound, leading towards an idea of that world as a representation of something else. It is enough to appreciate the intense listening with which poets such as Shelley, Keats, Wordsworth and Coleridge developed their writing as the external world, full of

---

[45] Ibid.
[46] P. B. Shelley, 'To a Skylark' in *Selections from Shelley's Poetry and Prose*, D. Welland (ed.) (London: Hutchinson Educational, 1961), p. 105.
[47] Ibid., pp. 105–6.

change, revolution and burgeoning industry, grew noisier and more insistent. In this context we see them as sound recordists of the highest order, writing poems as preserved soundscapes that teach the potential of deep attentiveness to some of the most subtle signals emerging from the space in which they lived and moved.

8

Honest John: The sound world of John Clare

While Wordsworth, Coleridge, Keats and others debated in verse the meanings of sound within their personal responses to the natural world, John Clare (1793–1864) seems to have been a kind of conduit through which the auditory signals of nature flowed unfiltered, directly onto the page. Clare was an instinctive naturalist who explored the soundscapes of the land and its inhabitants almost from the inside, with a genius for linguistic invention enabling him to articulate what many would consider to be unsayable. His home was the village of Helpston, geographically on the cusp of two landscapes, where Northamptonshire woods and fields open into the flat earth and wide skies of the eastern fenlands. His rural knowledge, learnt from being literally in the field, his sense of music (he was a fiddle player and folk song collector) and his ability to *speak on behalf* of nature, show him to be not only an acute listener, but also a witness who *understood* what he was hearing and its implications. In terms of literary expression of this, a door was opened for him when as a boy, in 1806 someone showed him a broken and tattered copy of James Thomson's *The Seasons* (We encountered Thomson in an earlier chapter with his poem 'To the Memory of Sir Isaac Newton'). It was Thomson's book-length poem, famous in its time, that pushed at the portal for Clare, as with these opening lines from 'Spring':

> Come, gentle Spring, ethereal mildness, come;
> And from the bosom of yon dropping cloud,
> While music wakes around, veil'd in a shower
> Of shadowing roses, on our plains descend.[1]

It was the perfect time in his life for the young teenager to encounter this work. He walked to nearby Stamford to buy the book and, such was his impatience to immerse himself, he climbed over the estate wall of Burghley House on the way

---

[1] J. Thomson, 'Spring' from The Seasons' in *The Poetical Works of James Thomson* (Edinburgh: William P. Nimmo, 1893), p. 3.

back to Helpston to devour the poem in full. Jonathan Bate, in his biography of the poet, quotes Clare as saying that 'what with reading the book and beholding the beauties of artful nature in the park, I got into a strain of descriptive rhyming on my journey home'.[2] This first composition, the first he committed to paper, was a piece called 'The Morning Walk'. Clare later destroyed a number of his first poems, but those that survive give us a sense of where his mature work would take him. It is significant that while enjoying the beauties of Burghley Park, he was aware that the landscape gardeners of the estate had created 'artful nature', a shaping of the natural world that had been reflected in pastoral verse, and that he would come to detest. One of his earliest preserved poems, here in Clare's own spelling and lack of punctuation, is 'Narrative Verses Written after an Excursion from Helpston to Burghley Park', beginning with a mature blend of sound and visual imagery:

> The faint sun tipt the rising ground
> No blustry wind – the air was still
> The blue mist thinly scatterd round
> Verg'd along the distant hill
> Delightful morn – from labour free
> I jocund met the southwest gale
> While here and there a busy bee
> Hum'd sweetly oer the flowery vale[3]

It is a substantial piece of work, continuing for another twenty stanzas, and was included in his 1821 collection, *The Village Minstrel*, but with a note from his publisher, John Taylor of Taylor and Hessey, that Clare had informed him of some early work in the book which was 'ten or twelve years old'.[4] Taking the older of these options, that could put the Burghley poem at around 1809, when Clare was sixteen, but it could have been earlier still; whatever the date, he may well have had Thomson's work in his mind, associated with the place in which he first became immersed in it. 'For Clare, as for Wordsworth (who regarded it as almost the only worthwhile nature poetry of the eighteenth century), *The Seasons* was memorable above all for its descriptions – of weather, of landscapes, of Nature

---

[2] J. Bate, *John Clare: A Biography* (London: Picador, 2003), p. 90.
[3] J. Clare, In R. Robinson, and D. Powell (eds) *The Early Poems of John Clare 1804–1822, Volume II* (Oxford: Clarendon Press, 1989), p. 4.
[4] Ibid., p. 783.

in all her varied moods and colours.'[5] It is not hard to sense the imagination of Clare being seized by passages such as this, from 'Winter':

> Assiduous in his bower, the wailing owl
> Plies his sad song. The cormorant on high
> Wheels from the deep, and screams along the land.
> Loud shrieks the soaring hern; and with wild wing
> The circling sea-fowl cleaves the flaky clouds.[6]

Thomson's use of language in this context was the kind of influence that would be reflected in Clare's own search for the exact sound to startle the eye and the ear in his work. It took an apprenticeship spent within the landscape, however, for those skills to develop in his own personal poetic identity. Robinson and Fitter have claimed that it was 'not until his middle period, the years between 1824 until 1832, [that] he demonstrated consistently his characteristic voice both as poet and naturalist'.[7] Indeed, as his writing took flight, it would not be long before it began to draw attention to itself. There had of course been 'peasant poets' before. Both Clare and Wordsworth admired the work of Robert Bloomfield (1766–1823), the Suffolk poet whose long poem *The Farmer's Boy* was also influenced by Thomson's *The Seasons*. We met Bloomfield – and Giles, the boy of the poem – in an earlier chapter, but his work deserves to be revisited here. Clare called Bloomfield 'the greatest Pastoral Poet England ever gave birth to,'[8] and *The Farmer's Boy*, published with woodcuts by Thomas Bewick in 1800, when Clare was just seven years old, like *The Seasons*, would have been formative for him. There were a number of parallels between the two poets, not least, the sensation caused by their appearance on the literary scene. *The Farmer's Boy* sold over 25,000 copies in the first two years after publication, it was translated into German, went through several American editions and was admired by and influenced the painter John Constable, as well as gaining praise from Wordsworth and Robert Southey. Like Clare, Bloomfield's latter days were surrounded by tragedy and poverty, but his mark on the time remains strong, and as John Lucas has written, 'we cannot hope properly to understand that historical period which is habitually called Romanticism if we do not pay

---

[5] J. Bate, *John Clare*, pp. 89–90.
[6] J. Thomson, 'Winter' from 'The Seasons.' *Poetical Works*, p. 160.
[7] R. Robinson and R. Fitter, *John Clare's Birds* (Oxford: Oxford University Press, 1982), p. vii.
[8] J. Bate, *John Clare*, p. 545.

attention to the works of Robert Bloomfield'.[9] In 'Summer', the second part of *The Farmer's Boy*, there is a passage in which listening goes where sight can no longer follow, as Giles, the farmer's boy of the poem's title, stops his work in the field, entranced by a sky-lark:

> What can unassisted vision do?
> What, but recoil where most it would pursue;
> His patient gaze but finish with a sigh,
> When music waking speaks the sky-lark nigh.
> Just starting from the corn, he cheerily sings,
> And trusts with conscious pride his downy wings[10]

The bird rises, and the boy tries to follow its progress with his eyes, but the sunlight dazzles him, the bird 'lost for awhile, yet pours her varied song' until, in an image that seems to pre-figure George Meredith's famous later poem, 'The Lark Ascending', it flies higher and still higher, until:

> E'en then the songster a mere speck became,
> Gliding like fancy's bubbles in a dream.[11]

Clare would have felt vindicated by knowledge of Bloomfield's story and inspired by his subject matter. As he established a poetic reputation himself, his work was taken on by Taylor and Hessey, also the publishers of John Keats, and for a time he was adopted by the fashionable London literati, ultimately finding himself isolated as a curiosity and novelty, as fashions changed and interest in his work faded. Keats and Clare never met, despite sharing a publisher, and it is ironic, as Christy Edwall wrote, that 'owing to the temporary fashion for "peasant poets", Clare's first collection, *Poems Descriptive of Rural Life and Scenery* (1820), vastly outsold Keats, although Keats got the better of Clare in the battle for canonicity'.[12] Equally, as Clare's reputation grew outside his own environment, he became torn between the society of literary London and his own often unschooled neighbours. Likewise, there was the requirement to feed and clothe his family, against his compulsion to write. He began to experience bouts of depression, and his alcohol consumption increased. He suffered delusions, believing himself

---

[9] J. Lucas, in J. Goodridge and J. Lucas (eds.) *Selected Poems of Robert Bloomfield* (Nottingham: Trent Editions, 1998), p. xxii.
[10] R. Bloomfield, ibid., p. 14.
[11] Ibid., p. 15.
[12] C. Edwall, 'A Few Beats of the Heart' in the *Times Literary Supplement*, 22 August, 2018. https://www.the-tls.co.uk/a-few-beats-of-the-heart/ (accessed 16 June 2022).

to be married to two women, Martha, and an idealized image of a childhood sweetheart, Mary Joyce. He was committed to a private asylum in Essex in 1837, and subsequently to Northampton General Lunatic Asylum, now St Andrew's Hospital, where he continued to write up until his death of a stroke at the age of seventy-one in 1864.

John Clare is one of the greatest poets of the English lyrical tradition, demonstrating in almost all his writing an acute and highly knowledgeable awareness of the natural world. In his work, we can hear the landscape of early to mid-nineteenth-century England, and in some cases, precise moments of everyday village life, as in an undated manuscript account of an ordinary morning in his village. Clare seldom used punctuation, but the sense of immediacy and place is palpable and moving in his writing. Coming through the window of his Helpston cottage he hears:

> The whistle of the ploughboy past the window making himself merry and trying to make the dull weather dance to a very pleasant tune which I know well and yet cannot recollect the song but there are hundreds of these pleasant tunes familiar to the plough and the splashing stream and the little fields of spring that have lain out the brown rest of winter and green into mirth with the sprouting grain the songs of the sky lark and the old songs and ballads that ever occupy field happiness in following the plough – [but] neither heard known or noticed by all the world beside.[13]

It is a vivid snapshot that in turn evokes a considered reflection by Clare on the nature of the extreme localness of rural society before the transport revolution that would change social mobility and the character of communities from the mid-nineteenth century onwards. Perhaps the most significant part of this lovely passage is the very last line: 'neither heard known or noticed by all the world beside'.

There is also the matter of his language, his word usage, often involving dialect, but most importantly, his practice of finding or inventing the right word or phrase to replicate with the utmost precision the sounds he was hearing. His writing of the Northamptonshire countryside around Helpston, six miles to the north of Peterborough, grew out of his first-hand experience in the fields, with jobs from childhood that kept him rooted to the land and to poverty. He had an honest irritation with what he saw as the genteel idealization of rural life

---

[13] G. Deacon, *John Clare and the Folk Tradition* (London: Sinclair Browne, 1983), p. 74.

conveyed in much of the poetry and prose that had created a park-like idea out of his world:

> Pastoral poems are full of nothing but the old thread bare epithets of "sweet singing cuckoo", "lovelorn nightingale" "fond turtles", "sparkling brooks", "green meadows", "leafy woods" etc etc. These make up the creation of Pastoral and descriptive poesy, and everything else is considered low and vulgar, in fact they are too rustic for the fashionable or prevailing system of rhyme.[14]

In an 1830 letter, he went so far as to directly criticize Keats for his portrayal of what Clare saw as 'nature as she … appeared in his fancys [*sic*] and not as he would have described her if he had witnessed the things he describes'.[15] The counter to this from Keats was that while liking Clare's poem, 'Solitude', he felt that '"the description too much prevailed over the sentiment"'. Be that as it may, we would not be without such images as 'the breeze, with feather-feet,' [goes] 'Crimping o'er the waters sweet'[16] or the silent night-sound of moonlight on a forest-bound ruin as it 'splinters through the broken glass'.[17] Nevertheless, both statements, coming from their authors, are significant and indicative. The reality of country life was hardship, poverty, and in Clare's time the industrialization of farming, enclosure and class tensions were factors of which he was strongly aware and about which he was vocal. Jeremy Hooker has linked the painter Constable, Wordsworth and Clare, who 'all, in different degrees, experienced strain, for the revolutions in industry and agriculture were pulling apart the world they would hold together in images of wholeness'.[18] Clare's prose, particularly in his letters, expresses the findings of a naturalist, and Gilbert White would have approved of and agreed with his description of the nightjar, to which he gave its country names of 'goatsucker' and 'fern-owl'. The poetry for him was in the sound of the creature, and the world it inhabited, and the precision of portrayal is enough to evoke the sound as an almost palpable thing, 'recording' the sonic experience and reflecting on its emotional impact:

---

[14] J. Clare, in R. Robinson, and G. Summerfield (eds.) *Selected Poems and Prose of John Clare* (Oxford: Oxford University Press, 1978), p. 66.

[15] Quoted in R. Mabey, 'Fantasists of the Fields?' in *Turning the Boat for Home* (London: Chatto and Windus), p. 219.

[16] J. Clare, 'Solitude' in J. W. Tibble (ed.) *The Poems of John Clare, Volume 1* (London: Dent, 1935), p. 195.

[17] Ibid., p. 197.

[18] J. Hooker, 'The Tree of Life: Explorations of an Image' In *Art of Seeing: Essays on Poetry, Landscape Painting and Photography* (Swindon: Shearsman Books, 2020), p. 102.

> They [nightjars] make an odd noise in the evening, beginning at dewfall and continuing it at intervals all night. It is a beautiful object in poetic Nature – *(nay all nature is poetic)* [My Italics]. From that peculiarity alone one cannot pass over a wild heath in a summer evening without being stopped to listen and admire its novel and pleasing noise. It is a trembling sort of crooing sound which may be nearly imitated by making a crooing noise and at the same time patting the finger before the mouth to break the sound, like stopping the hole in a German flute to quaver a double sound on one note. This noise is generally made as it descends from a bush or tree for its prey.[19]

There is the musician talking, and it is clear that he has rehearsed the sound himself, such is his attention to detail. The bird fascinated Clare; in an account stretching over two pages, he describes it in detail, including anecdotes of where he first heard it 'on my love rambles' when courting his future wife, Martha. Towards the end of the piece, he returns to the sound, in a further attempt at capturing it precisely:

> It very often startled me with its odd noise which was a dead thin whistling sort of sound, which I fancied was the whistle call of robbers, for it was much like the sound of a man whistling in fear of being heard by any of his companions, tho [*sic*] it was continued much longer than a man could hold his breath. It had no trembling in it like a gamekeeper's dog-whistle but was of one long continuous sound.[20]

As Peter Levi has observed, 'nothing is locked away from anyone; he wants to teach, he wants to tell. Nature to him was the proper subject of poetry ... [and] by nature he meant something that could be very exactly studied and very personally recorded.'[21] No wonder he was out of patience with what he saw as the metaphysical fantasies of suburban city-dwellers. Clare's vast output of poems continued into his asylum years, when increasingly they became mnemonics, with birdsong playing in his mind as a consolation and a recollection. Yet throughout his writings, there is an immediacy of sound, image and thought caught in the field, exactly expressed unembellished and without sentiment:

> The frog croaks loud, and maidens dare not pass
> But fear the noisome toad and shun the grass;
> And on the sunny banks they dare not go

---

[19] J. Clare quoted in R. Robinson and R. Fitter, pp. 65–6.
[20] Ibid., pp. 66–7.
[21] P. Levi, *John Clare: Bird Poems* (London: Folio Society, 1980), p. 16.

> Where hissing snakes run to the flood below.
> The nuthatch noises loud in wood and wild,
> Like women turning skreeking to a child ... & c.[22]

He expressed his frustration at fashionable shortcuts to familiar birdsong at the start of his short poem on 'The Wren':

> Why is the cuckoo's melody preferred,
>   And the nightingale's rich songs so madly praised
> In poets' rhymes? Is there no other bird
>   Of nature's minstrelsy, that oft hath raised
> One's heart to ecstasy and mirth as well?[23]

This is not to say that the more famous birds are not justly acclaimed in Clare's poetry. Quite the opposite in fact: he is only calling for equality and recognition here. He is also pointing to the fact that there are layers of bird sound that require an understanding that some writers are not prepared to explore. He himself wrote more than once of the nightingale, and in his great poem 'The Nightingale's Nest', which Michael Longley has called 'among the glories of English poetry',[24] he shows how closely he is able to observe both the sound and the bird itself. A skilled and loving birdwatcher, with a true understanding, he is a guide to us, as if we were standing beside him:

> For here I've heard her many a merry year
> At morn and eve, nay all the live long day
> As though she lived on song – this very spot.[25]

He remarks on the juxtaposition between the beauty of the song, and the somewhat unspectacular visual appearance of the bird, while at the same time observing how the little creature puts its whole physical being into producing its music:

>   I've nestled down
>   And watched her while she sung – and her renown
> Hath made me marvel that so famed a bird

---

[22] J. Clare from 'Dyke Side', ibid., p. 114.
[23] J. Clare. 'The Wren', ibid., p. 101.
[24] M. Longley, 'A Note on John Clare' in *Sidelines: Selected Prose 1962-2015* (London: Enitharmon Press, 2017), p. 116.
[25] J. Clare, 'The Nightingale's Nest' in E. Robinson and G. Summerfield (eds) *Clare: Selected Poems and Prose* (Oxford: Oxford University Press, 1975), p. 119.

> Should have no better dress than russet brown.
> Her wings would tremble in her extacy [sic]
> And feathers stand on end as twere with joy,
> And mouth wide open to release her heart
> Of its out-sobbing joys[26]

Was ever there a more moving and personal response to sound? And who else but Clare would think of joys as 'out-sobbing'? As Longley writes, 'he saw the natural world with such preternatural clarity that exultation and anguish often combine in his writing'.[27] It is true; towards the end of this extraordinary poem, he turns from us to address the nightingale itself, as the song fills the woodland and seems to be absorbed like perfume into the surfaces of everything it touches:

> For melody seems hid in every flower
> That blossoms near thy home. These harebells all
> Seem bowing with the beautiful in song;
> And gaping orchis, with its spotted leaves
> Seems blushing of the singing it has heard[28]

John Clare teaches us what to hear, how to listen to it and then the wonder of it: 'Under his devout gaze humdrum objects are transfigured, astonishment is always near at hand'.[29] In his poem, 'The Progress of Rhyme' the nightingale is the subject of his most ambitious attempt to capture birdsong in mimetic terms. This long poem is full of sound caught in increasingly vivid ways as he chronicles how 'The bird or bee its chords would sound,/The air hummed melodies around',[30] each bringing its own joy to his ear, so that 'e'en the sparrow's chirp to me/Was song in its felicity'.[31] In this we gain a clear indication that Clare's method was to commit impressions directly to the page *in situ*, to give us an analogy to today's sound recordist, to capture the moment on a memory card. J. W. Tibble, in his introduction to the two-volume 1935 edition of Clare's poems, points out:

> Like Wordsworth's, his poems are usually the record of an experience already completely realized, but unlike Wordsworth he seldom allowed an interval

---

[26] Ibid.
[27] M. Longley, 'A Note on John Clare', in *Sidelines: Selected Prose 1962–2015* (Enitharmon Press, 2017), p. 116.
[28] J. Clare, *Clare: Selected Poems and Prose*, p. 121.
[29] Longley, 'A Note on John Clare', p. 116.
[30] J. Clare, 'The Progress of Rhyme' in Tibble, volume 1, p. 436.
[31] Ibid., p. 437.

to elapse between the experience and its recording. 'I found the poems in the fields', he said, and often enough he wrote them in the fields too, on any scrap of paper that came to hand. Like Keats, he experienced the immediate excitement of composition; but the poem did not arise primarily out of that excitement. It was not the words that excited him, the actual making of the poem, so much as the power of the words to prolong and renew that rapture which he felt in the presence of nature.'[32]

While we have seen and 'heard' attempts to preserve the experience of sound through text across many centuries, Clare seems to be a 'modern', a contemporary of ours in his desire to hold the moment with exact precision. In the progress and refinement of his nature writing, we have an analogy with the late twentieth century's developing technical innovations regarding the quality of sound recording, the move over a relatively few years, from cylinder to disc to tape and ultimately to digital sound cards and the sophistication of microphones capable of capturing pure sonic moments and preserving them from the fading of time. John Clare's work provides a staging post, a movement forward in the directness of recording and expression. He was a field sound recordist of his time; it was just that he was two hundred years or so ahead of the technology. Indeed, he had no need of it, because he possessed his own in-built sound memory card: 'and each old leaning shielding tree/ Were princely palaces to me,/Where I would sit me down and chime/my unheard rhapsodies to rhyme.' So the outpouring continues until we come to the nightingale itself:

> The more I listened and the more
> Each note seemed sweeter than before,
> And aye so different was the strain
> She'd scarce repeat the note again:
> 'Chew-chew chew-chew,' and higher still:
> 'Cheer-cheer cheer-cheer,'more loud and shrill:
> 'Cheer-up cheer-up cheer-up,' and dropt
> Low: 'tweet tweet jug jug jug,' and stopt
> One moment just to drink the sound
> Her music made, and then a round
> Of stranger witching notes was heard,
> As if it was a stranger bird:
> 'wew-wew wew-wew, chur-chur chur-chur,

---

[32] J. W. Tibble, ibid., Introduction, p. viii.

> Woo-it woo-it': could this be her?
> 'Tee-rew tee-rew tee-rew tee-rew,
> Chew-rit chew-rit,' and never new:
> 'Will-will will-will, grig-grig grig-grig.'
> The boy stopt sudden on the brig
> To hear the 'tweet tweet tweet' so shrill,
> Then 'jug jug jug' and all was still ... & c.[33]

It is remarkable to have so faithful a recording contained within a text without disturbing the natural and logical flow of rhyme and rhythm. The whole poem unfolds with the spontaneity of ... well, birdsong! As seen here, and throughout his work, Clare does not limit himself to human language's conventional meanings if it can be appropriated in a new way to 'syllable the sound' of nature. In his work, poetic language takes up the mantle of audible nature. The poem, 'Flight of Birds', for example, provides every aspect of bird sound, even beyond song itself:

> The crow goes flopping on from wood to wood,
> The wild duck wherries to the distant flood
> The pigeon suthers by on rapid wing
> Whizz goes the pewit o'er the ploughman's team,
> With many a whew and whirl and sudden scream .... & c.

Beyond the particulars, Clare can use a wide-angle, stereo soundscape to convey the environment. His poems are writings of a place in which things happen, and those happenings may be climatic, bird- or animal-related or simply the sounds of the place being itself, the world, as it were, going on around him. One of the regular interlopers into the sound of the landscape in Clare's day would have been village bells, of which Alain Corbin has written: 'In the nineteenth century, at least in the countryside, bell ringing defined a space in which only fragmented, discontinuous noises were heard ... After all there were no airplanes, which nowadays are capable of competing with, overwhelming and, above all, *neutralising* the sound of bells.'[34] The sonic perspective of ringing bells across a landscape, say on a Sunday morning coming over the flat lands to the east of Helpston, from villages at varying distances, would have been a call to prayer and conscience. To a modern ear, bells across the meadow might well enhance

---

[33] Ibid., p. 439.
[34] A. Corbin, 'Identity, Bells and the Nineteenth Century French Village' in M. M. Smith (ed.) *Hearing History: a Reader* (Athens/London: University of Georgia Press, 2004), p. 185.

the rural ambience, but Clare's church was mostly in the fields and hedgerows, and often when church bells intrude into his field of sound, he responds on behalf of a particular set of social values, devout perhaps, but with their own priorities. 'Sunday' for example:

> The bell, when knoll'd its summons once and twice,
> Now chimes in concert, calling all to prayers;
> The rustic boy who hankers after vice
> And of religion little knows or cares
> Scrambs up his marbles, and by force repairs,
> Though dallying on till the last bell has rung.[35]

He was an activist with his pen against enclosure, railing against artificiality, pretence and hypocrisy, and he wrote scathingly of the gentrification of the countryside, speaking of 'pride and fashion...creeping out of the citys [sic.] like a plague to infest the Village.'[36] While the lad in the poem is church-bound under sufferance, there are others Clare knew for whom formal religion is both a luxury and an irrelevance, as he reminds us in 'A Sunday with Shepherds and Herdboys', here in his unpunctuated version:

> The shepherds and the herding swains
> Keep their sabbath on the plains
> They know no difference in its cares
> Save that all toil has ceas'd but theirs
> For them the church bells vainly call
> Fields are their church and house and all
> Till night returns their homeward track
> When soon morns suns recall them back[37]

These are figures in a landscape, but Clare had learnt early from Thomson and others that the landscape itself was its own subject; how could it not be, since as he grew and lived, he changed from being an observer to a participant in the natural world, finally to the point of immersion wherein there was a oneness, an indivisible partnership between him and his environment. He was not operating a recording device: he WAS the device. But he supplemented this direct communication with his reading, and a long list of writers he admired about

---

[35] J. Clare in Tibble, l, p. 189.
[36] Ibid., p. 73.
[37] J. Clare, in *Selected Poems and Prose*, pp. 134–5.

nature was included in a letter to his publishers, written in 1822, showing wide-ranging influences, including Spencer, Cowley, Shakespeare, 'the Elizabethan poets of glorious memory' Gay, Wordsworth, Bloomfield and many others. He adds that 'I always feel delighted when an object in nature brings up in ones mind an image of poetry that describes it from some favourite author … To look on nature with a poetic eye magnifys [sic] the pleasure she herself being the very essence and soul of Poesy.'[38]

He also names Milton among his favourites, and when he does so, he is specific, rather than as with other writers, simply listing; here, he cites two poems in particular, 'L'Allegro' and 'Il Penseroso' and interestingly, we have a direct witness to a Wordsworthian connection to the very same poems in Dorothy Wordsworth's Grasmere Journal for Christmas Eve, 1802, in which she offers some intimate domestic context:

> William is now sitting by me at half past ten o'clock. I have been beside him ever since tea running the heel of a stocking, repeating some of his sonnets to him, listening to his own repeating, reading some of Milton's 'Allegro' and 'Penseroso'. It is a quiet keen frost. Mary is in the parlour below attending to the baking of cakes and Jenny Fletcher's pies.[39]

The two poems are indeed linked; 'L'Allegro' is Italian for 'the cheerful man', while 'Il Penseroso' is Italian for 'the contemplative man'. Furthermore, there are passages within the poems which we may easily understand would have appeal to the sensibilities of both Wordsworth and Clare, offering as they do images of an auditory sense of place, beginning in specific particulars and widening into a landscape, as here, in 'L'Allegro'

> While the cock with lively din,
> Scatters the rear of darkness thin,
> And to the stack, or the barn door,
> Stoutly struts his dames before,
> Oft list'ning how the hounds and horn
> Cheerily rouse the slumb'ring morn.[40]

and here, in 'Il Penseroso':

---

[38] Ibid., pp. 56–7.
[39] D. Wordsworth, *The Grasmere Journal* (Oxford: Oxford University Press, 2008), pp. 134–5.
[40] J. Milton, "L'Allegro' In J. Leonard (ed.) *John Milton: The Complete Poems* (London: Penguin Books, 1998), p .23.

> Oft on a plat of rising ground,
> I hear the far-off curfew sound,
> Over some wide-watered shore,
> Swinging low with sullen roar.[41]

It is not hard to see how Clare found vindication for his work in poetry like this. Mostly, however, it was from the Place itself, a link that was absorbed into his mind and thence onto the page. His prose, like his verse, gains its astonishing presence, immediacy, and precision from the spontaneous outpouring through him onto the page. Wordsworth may have 'recollected in tranquillity' but in John Clare's poetry and prose, the sounds and spirit of nature are using him as a transmitter right there in the field:

> March 25th, 1825: I took a walk today to botanise and found that the spring had taken up her dwelling in good earnest … the sallows are cloathed [sic] in their golden palms where the bees are singing a busy welcome to spring.'[42]

Today, our microphone may record the sound of bees, but it takes an attentive mind listening, to be aware of their song. We have encountered Clare's use of dialect words already, as well as sounds formed into syllables of his own invention, and he makes no concessions in his prose when it comes to finding the exact sound equivalent in writing:

> The wood pigeons are busily fluskering[43] among the ivied dotterels[44] on the skirts of the common … Have you never heard that cronking[45] jarring noise in the woods at this early season? I heard it today and went into the woods to examine what thing it was that caused the sound and I discovered that it was the common green woodpecker busily employed at boring his hole, which he effected by twisting his bill round in the way that a carpenter twists his wimble,[46] with this difference, that when he got it to a certain extent he turns it back and then pecks awhile, and then twists again. His beak seems to serve all the purposes of a nail passer,[47] gough[48] and wimble effectually.[49]

---

[41] J. Milton, 'Il Penseroso', Ibid., p. 27.
[42] J. Clare, *Selected Poems and Prose*, p. 153.
[43] To flusker: To fly with sudden motion, with an element of noise and commotion.
[44] Dotterel: A pollard tree, what Clare called 'old stumping trees in hedge rows, that are headed or lopped every ten or twelve years for fire-wood' (ibid., p. 241.)
[45] To cronk: To croak or honk, used especially of frogs and geese.
[46] Wimble: A gimlet or auger.
[47] Nail passer: Corruption of nail-piecer, another word for gimlet or auger.
[48] To gough: To gouge.
[49] J. Clare, *Selected Poetry and Prose*, p. 154.

Rooted in a place, his work does not require us to be there to know it. We listen to Clare's 'recordings', and wherever we are, the sound and the spirit of them communicates. This whole passage is a sound/natural history workshop, and the wonder of it, mostly, is not in its portrayal of a specific woodland, (although it conveys that too), but in the universality of its observation. As Erin Lafford has written:

> Clare is valued as a poet of direct communication. His poems are filled with Northamptonshire dialect that fosters an instantaneous connection to his local environment, creating an immediate sense of place through sound. Likewise, Clare's representations of natural sounds, such as the 'whewing' of the pewit, the 'swop' of the jay bird as it flies, and the 'chickering crickets', have a mimetic quality that creates a direct experience of what he hears.[50]

This does not mean that the localness of his work makes for limitation; on the contrary, he is a sounding board and interpreter on behalf of a place, to the wider world. Equally, as he 'took a walk to botanise' on 25 March 1825, the words he has left us from that day have the ring of notetaking, as much for himself as for any external reader. Lafford writes of 'Clare's mutterings, murmurings and ramblings', taking her meaning of 'mutterings' from the OED as to 'grumble in an undertone', 'to make a low ominous rumbling sound' and 'to recite in low indistinct tones'. Having her definition, she points out that what these expressions have in common is

> The way in which muttering lingers at the boundary between language and sound. All the above descriptions fall into use of the word 'tone' or 'sound' as well as 'speak', 'express', and 'recite'. To mutter is therefore to speak a sound, or rather to utilise sounds in a different way to standardised speech. Muttering does not necessarily invite others to hear and understand what is being said.[51]

Taking this into consideration, we might think that Clare's direct voice is as much to himself as to an outsider, indeed perhaps the voice of a person undergoing mental health issues, (as he increasingly was). While this would be on the face of it to negate his abilities and desires as a teacher and communicator, there is an underlying truth that suggests it is through his pure honesty of expression, that *his* sound 'broadcasts' his *subject's* sound in the highest definition possible.

---

[50] E. Lafford, 'Clare's Mutterings, Murmurings, and Ramblings: The Sounds of Health.' Accessed May 2022. https://core.ac.uk/reader/200747870.
[51] Ibid.

To say that when 'on location' (as he almost always is), Clare is absorbed in his sense of walking through a landscape would be to miss the point, because Clare, as he focuses and concentrates his lens and microphone, becomes increasingly absorbed *within* the landscape until he and the details he observes are as one. John Barrell has suggested that Clare 'does not detach himself from the landscape … or post himself on a commanding height but describes only what is immediately around him. The attempt, then, is not so much to describe the landscape, or even to *describe* each place, as to suggest what it is like to be in each place.'[52] This is the overall purpose of text as sound recorder too, that is to say not only to evoke an acute sense of the sound of a place, being or thing, but to preserve the feeling of it, or rather the *experience* of feeling it, in Barrell's eloquent phrase, 'to suggest what it is like to be in each place.'

Clare, because he is a naturalist, knows where to find his birds, and while he has the same ability as Ludwig Koch when it comes to isolating a particular sound, his writing provides us with the context of the place in which he locates it, and in this his precision of focus is no less sharp than his attention to the particular. It is not his way to offer us a vague water colour wash of general wild track in his field recordings, but a sharp, vivid image of stereophonic text. Before circumstances took him away, the local landscape of his life provided a continuity upon which he was able to reflect. These were not places once seen and then abandoned, but corners and cupboards of experience to which he returned, containing intimate and endless sources of comparative observation. Such a place was Emmonsail's Heath, not far from his home. Clare knew every inch of it, its flora and fauna, and indeed many of its human itinerant inhabitants, including the travelling people who passed through and sometimes spent nights there; it was a place he lived and breathed. He could come to a much loved and familiar place such as this and write directly to it, as to a person: 'In thy wild garb of other times/I find thee lingering still,'[53] or recall it from memory as a fundamental in his life, a yardstick enabling him to take comfort from its seemingly unchanging world with which he could commune as with a friend: 'All the same old things to be/ As they have ever been.'

> The brook o'er such neglected ground,
> One's weariness to soothe,

---

[52] J. Barrell, *The Idea of Landscape and the Sense of Place, 1790–1840* (Cambridge: Cambridge University Press, 1972), pp. 110–13.

[53] J. Clare, 'Emmonsale's [sic] Heath' in Tibble, volume I, p. 382.

> Still wildly winds its lawless bound
> And chafes the pebble smooth[54]

Then, having given us a wide-angle soundscape, gradually homing in on the stream, in a later and much-admired sonnet, 'Emmonsail's Heath in Winter'[55] he directs his listening apparatus towards the wildlife itself:

> While the old heron from the lonely lake
> Starts slow and flaps his melancholy wing
> And oddling crow in idle motions swing
> On the half-rotten ash-tree's topmost twig.[56]

Even when he became ill in his later years, his output was always prodigious, and although sometimes losing the focus of his earlier work, he remained able to summon – often from poignant memory – sights and particularly sounds from past days, as in his 'Pleasant Sounds', hovering somewhere between prose and poetry, and written in Northampton Asylum. It is as though he is playing back a tape in his mind, re-running the sounds of his own being:

> The rustling of leaves under the feet in woods and under hedges;
> The crumping of cat-ice and snow down wood-rides, narrow lanes, and every street causeway;
> Rustling through a wood or rather rushing, while the wind halloos in the oak-top like thunder;
> The rustle of birds' wings startled from their nests or flying unseen into the bushes;
> The whizzing of larger birds overheard in a wood, such as crows, puddocks, buzzards;
> The trample of robins and woodlarks on the brown leaves, and the patter of squirrels on the green moss;
> The fall of an acorn on the ground, the pattering of nuts on the hazel branches as they fall from ripeness;
> The flirt of the groundlark's wing from the stubbles – how sweet such pictures on dewy mornings, when the dew flashes from its brown feathers![57]

Late as it was in his life, ill as he was, still there are the moments here that startle with their exactness and sense of wonder; for example, how perfect is the sound

---

[54] Ibid., p. 383.
[55] Clare's spelling of the place varies.
[56] J. Clare, Tibble, volume II, p. 146.
[57] J. Clare, Tibble, volume II, p. 427.

of the 'flirt of a groundlark's wing'? And how acute must be the sensitivity of a personal inner sound recorder to register the fall of an acorn, or a robin's movement over leaves as a 'trample'? The wild tracks of Clare's homeland played on in his mind even beyond the time when he was able to express the sights and sounds in his greatest poetry. On 26 February 1848, he wrote to his son Charles from Northampton asylum, offering some fatherly advice 'to study mathematics, astronomy, languages and botany as the best amusements for instruction'. He also suggests angling as a recreation: 'I was fond of it myself and there is no harm in it if your taste is the same – for in those things I have often broke the Sabbath when a boy and perhaps it was better then [sic] keeping it in the village hearing scandal and learning tipplers frothy conversation. "The fields his study, nature was his book".[58] It is a poignant and revealing letter, with flashes in phrases that move him from thought to thought, opening recollections that have a touching immediacy about them: 'I wrote or rather thought poems' he says, remembering that once his creations were things that 'men read and admired:

> I loved nature and painted her both in words and colours better than many poets and painters … In my boyhood Solitude was the most talkative vision I met with. Birds, bees, trees, flowers all talked to me incessantly louder than the busy hum of men … Who so wise as nature out of doors on the green grass by woods and streams under the beautifull [sic] sunny sky – daily communings with God and not a word spoken.[59]

A late sonnet, dated 10 February 1860, just over four years before his death, and one of the last six to survive in his own hand, begins: 'Well, honest John, how fare you now at home?/ The spring is come, and birds are building nests.'[60] It is a simple poem of domestic life, surrounded by the natural world that had been a part of that world for him, and from which at the time of its writing, he had been separated for decades. Yet the idea of that beloved landscape and its sounds played on in his mind, and the poems – somewhat varied and inconsistent as they became in the latter years – continued to hum and whisper with the music to which he had been such an intent listener, and the soundscape of nature that he had so long orchestrated in his purest work.

---

[58] J. Clare, 'Letter to Charles Clare' in M. Storey (ed.) *John Clare: Selected Letters* (Oxford: Oxford University Press, 1990), p. 209.
[59] Ibid., pp. 209–10.
[60] J. Clare, Tibble, volume II, p. 518.

9

North American sublime

John Clare found his poems in the fields, and looked askance at town dwellers who wrote of nature as he sometimes considered, through fanciful imaginations, from more circumscribed physical parameters. Yet there were those who barely set foot beyond the confines of a garden for whom the world of nature spoke with a profound and eloquent voice. In about 1865, the year after Clare's death in Northampton Asylum, Emily Dickinson (1830–86), then in her mid-thirties and living at home in Amherst, Massachusetts, wrote:

> I never saw a Moor –
> I never saw the Sea –
> Yet I know how the Heather looks
> And what a Billow be.[1]

Dickinson spent much of her life in social isolation, and only ten of her 1,800 poems were published in her lifetime. Her brief, spare verses have latterly gained her a prodigious global reputation, and her almost epigrammatic work has the capacity to focus like a laser on an event, a subject, a sound or an emotion through intense minute observation and emotional self-examination. A key moment in her life – and that of literary history – was 15 April 1862, when the writer Thomas Wentworth Higginson received a cache of her work, together with a letter, enquiring of him whether he felt her work 'breathed'? She was at this time thirty-two, and Higginson found it to be unclassifiable; no one had created or read anything like it before. Much has been written of this work since 1955, when the first complete, mostly unaltered collection of her poetry was published under the editorship of Thomas H. Johnson. Dickinson is another landmark in our exploration because although her writing encompassed horizons no wider than

---

[1] E. Dickinson, 'I Never Saw a Moor' (1052) in T. H. Johnson (ed.). *The Complete Poems of Emily Dickinson* (London: Faber and Faber, 1977), p. 480. (Note: The numbering after the title refers to the cataloguing in the Johnson edition.)

her garden or occasionally nearby woods, her awareness of natural sounds and her ability to capture them in her texts has the capacity to startle the senses like a new song. Her letters to Higginson, among the most important correspondence in American and even world literature, often capture the extraordinary turns of phrase that are to be found in the poems. Sometimes these phrases jump out of an otherwise conventional paragraph, as here, from July 1862, shortly after their friendship began: 'I think you would like the chestnut-nut tree I met on my walk. It hit my notice suddenly, and I thought the skies were in blossom. Then there's a noiseless noise in the orchard that I let people hear.'[2] At that time, the American Civil War was sixteen months into its course, while far away that very month, in England, an Oxford don by the name of Charles Lutwidge Dodgson extemporized a story while on a rowing trip along the Isis in Oxford, for a girl called Alice Liddell. The publication of *Alice in Wonderland* by Lewis Carroll in 1865 was to open up surreal vistas of silent sound for readers overcoming generations. In the meantime, as time passed and the world beyond the homestead in Amherst went on, for Emily it was the internal effect of the minutiae around her that provided the subject matter. Almost ten years later, in a letter dating from the winter of 1871, she writes of the changing seasons with an eccentric perfection, evoking silence that begets an inner sound: 'When I saw you last, it was mighty summer – now the grass is glass, and the meadow stucco, and "still waters" in the pool where the frog drinks. These behaviours of the year hurt almost like music, shifting when it eases us most.'[3] An elliptical sentence like that can tease the mind as much as a whole page of poetry. When we come to the poems themselves, there is an implicit sound evoked on the page by her unusual punctuation:

> Great Streets of silence led away
> To Neighbourhoods of Pause –
> Here was no Notice – no Dissent
> No Universe – no Laws –
>
> By Clocks, 'twas Morning, and for Night
> The Bells at Distance called –
> But Epoch had no basis here
> For Period exhaled.[4]

---

[2] E. Dickinson, Letter to Thomas Wentworth Higginson, July 1862. In E. Fragos (ed.) *Emily Dickinson: Letters* (New York: Alfred A. Knopf, 2011), p. 172.
[3] Ibid., p. 180.
[4] E. Dickinson, 'Great Streets of Silence Led Away' (1159) in *The Complete Poems*, p. 517.

Her use of capital letters is one thing, but another is the stylistic employment of the dashes, something perhaps a little more than a comma, almost like a breath being taken, but certainly to be observed and interpreted and part of the overall sound of these minimal yet profound poems. Whatever the intention in her mind at the time of composition, for us they have the effect of slowing the reading – even a silent reading – and we pause, dwell on the lines, allow them to linger, a new sound-thought contemplated. Her physical horizons were intensely local, but her mental and imaginative world knew no boundaries, so she could write potently of the *idea* of things such as the sea, and the relationship between elements, through a kind of inner seeing and hearing:

> I think that the Root of the Wind is Water –
> It would not sound so deep
> Were it a Firmamental Product –
> Airs no Oceans keep –
> Mediterranean intonations –
> To a Current's Ear –
> There is a maritime conviction
> In the Atmosphere –[5]

It is, however, when we turn to Dickinson's own physical world that we see the depth of her knowledge both of natural history, and her precise eye and ear, at their greatest stretch and intensity. She was an avid gardener at the Amherst homestead, tending a small glass conservatory and a flower garden, and observing at close quarters the changing seasons within the microcosm of a fenced plot of land in which she walked with her dog, Carlo. She had studied botany at Amherst Academy and Mount Holyoke seminary, but it was the garden of 'The Homestead' on Main Street that Emily cultivated and to which she listened, as here in 1862, observing a hummingbird:

> Within my Garden, rides a Bird
> Upon a single Wheel –
> Whose spokes a dizzy Music make
> As 'twere a travelling Mill –
>
> He never stops, but slackens
> Above the Ripest Rose –
> Partakes without alighting
> And praises as he goes … (& c.)[6]

---

[5] Ibid., 'I think that the Root of the Wind Is Water' (1302), p. 567.
[6] Ibid., 'Within My Garden, Rides a Bird' (500), pp. 242–3.

There was a communion for her within the natural world that she did not find in organized religion; a poem of 1860 expresses her reasons for abstinence from church habits, and as a dissenter, the sources of her personal beliefs:

> Some keep the Sabbath going to Church –
> I keep it, staying at Home –
> With a Bobolink for a Chorister –
> And an Orchard for a Dome –
>
> Some keep the Sabbath in Surplice –
> I just wear my Wings –
> And instead of tolling the Bell, for Church –
> Our little Sexton – sings.
>
> God preaches, a noted Clergyman –
> And the sermon is never long,
> So instead of getting to Heaven, at last –
> I'm going, all along.[7]

The bobolink was a bird which wintered in southern parts of South America, notably in Paraguay, Argentina and Bolivia, but summered and bred in her part of North America. It was a small creature of which she wrote often. Since her time, its populations have rapidly declined due to habitat loss, and in recent years it has been declared a threatened species. Here in this poem she picks up on its clerical plumage, black, with a yellow nape and white scapulars, and it is not hard to imagine its plumage as the vestments of a religious officiate. Indeed, natural sounds were a consolation, a prayer book, a hymnal and a sermon for Dickinson. There was also, even in the smallest sounds, an almost unbearable mnemonic as she grew older and lost loved ones. The continuity of recurring seasons and all that came with them, the round of the year, the familiarity of birdsong as it returned in spring, formed a poignant counterpoint when set against human mortality:

> The saddest noise, the sweetest noise,
> The maddest noise that grows, –
> The birds, they make it in the spring,
> At night's delicious close

---

[7] Ibid., 'Some Keep the Sabbath Going to Church' (324), pp. 153–4.

The poem ends:

> An ear can break a human heart
> As quickly as a spear,
> We wish the ear had not a heart
> So dangerously near.[8]

Beyond the heart that hears through the ear, she writes elsewhere of a terrible inner stillness, a silent spring of the soul:

> There is no silence in the Earth – so silent
> As that endured
> Which uttered, would discourage Nature
> And haunt the World.[9]

In the end, the smallest sounds persist, a continuity of minutiae within – and emphasizing silence, reflected in this well-known poem:

> I heard a Fly buzz – when I died –
> The Stillness in the Room
> Was like the Stillness in the Air
> Between the Heaves of Storm (& c.)[10]

The soundscapes of Emily Dickinson reflect the world through which she moved, an inner space in which the hypersensitivity of her senses and her emotional responses had the capacity to absorb the tiniest visual and auditory clues around her, and interpret them in a totally unique way. When the microphone was eventually invented, it was the 'micro' in it that was deemed significant, seen as it was as a device by which sounds beyond the capacity of the human senses could be captured. Dickinson's tuning gave her just such a facility, and her poems are the recordings she left as evidence of this.

To step from Emily's garden and the tiny sounds of her birds and whispering breezes, to the epic landscapes through which John Muir (1838–1914) moved and worked is to risk a feeling of near agoraphobia. Muir came to Yosemite in Nevada firstly as a visitor in the late 1860s, and subsequently built a cabin there in which he lived and studied the terrain at close quarters for two years. Here is Muir, describing the sound of the Yosemite Falls:

---

[8] Ibid., 'The Saddest Noise, the Sweetest Noise' (1764), pp. 713–14.
[9] Ibid., 'There Is No Silence in the Earth – So Silent' (1004), pp. 465–6.
[10] Ibid., 'I Heard a Fly buzz – when I died –' (465), p. 223.

> This noble fall has far the richest, as well as the most powerful, voice of all the falls of the valley, its tones varying from the sharp hiss and rustle of the wind in the glossy leaves of the live-oak and the soft, sifting, hushing tones of the pines, to the loudest rush and roar of storm winds and thunder among the crags of the summit peaks. The low bass, booming, reverberating tones, heard under favourable circumstances five or six miles away are formed by the dashing and exploding of heavy masses mixed with air upon two projecting ledges on the face of the cliff, the one on which we are standing and another about 200 feet above it.[11]

John Muir was born in Dunbar, Scotland, but he emigrated with his father to the United States in 1849. He developed a love affair with America's great wild spaces that would last a lifetime, and today he is still revered as the father of the National Parks, and a pioneer of the conservation movement. He was many things in his lifetime: an inventor, a mountaineer, an explorer, a botanist, a geologist and encompassing it all, a born writer with a consummate skill for conveying wonder. Muir wrote much about the High Sierra, the Mountains of California, and Yellowstone and General Grant National Parks, as well as specific observations on their flora and fauna, including essays on the Water Ouzel, the Douglas Squirrel and Wild Sheep. During the spring of 1869, he took the job of a sheep herder on a ranch in the Sierra Nevada, as a means to an end for the funding of his explorations through California's Central Valley. As he explored, his growing familiarity with the place enabled him to become proficient enough to act as a guide, and some of Yosemite's most famous visitors followed him on his treks, including, among others, the writer Ralph Waldo Emerson. All these experiences were added to Muir's documentation, which culminated in the publication of *The Yosemite* in 1912. His witness of being present during a night-time earthquake in Yosemite remains an encapsulation of the awe the painters in America's 'sublime' movement sought to capture: even a sound recording of the event would be hard-pressed to surpass Muir's account. It is pictorial, sonic, poetic and conveys completely, the sense of being at the centre of a giant seismic event. And it begins in near silence:

> It was a calm moonlight night, and no sound was heard for the first minute or so, save low, muffled, underground, bubbling rumblings, and the whispering and rustling of the agitated trees, as if Nature were holding her breath. Then,

---

[11] J. Muir, *Journeys in the Wilderness: A John Muir Reader* (Edinburgh: Birlinn, 2017), pp. 444–5.

suddenly, out of the strange silence and strange motion there came a tremendous roar. The Eagle Rock on the south wall, about half a mile up the Valley, gave way, and I saw it falling in thousands of the great boulders I had so long been studying, pouring to the Valley floor in a free curve luminous from friction, making a terrible sublime spectacle – an arc of glowing, passionate fire, fifteen hundred feet in span, as true in form and as serene in beauty as a rainbow in the midst of the stupendous, roaring rock-storm. The sound was so tremendously deep and broad and earnest, the whole earth like a living creature seemed to have at last found a voice and be calling to her sister planets. In trying to tell something of the size of this awful sound it seems to me that if all the thunder of all the storms ever heard were condensed into one roar it would not equal this rock-roar at the birth of a mountain talus. Think, then, of the roar that arose to heaven at the simultaneous birth of all the thousands of ancient canyon-taluses throughout the length and breadth of the Range![12]

Here is the eye and ear witness of a cataclysmic happening emerging from stillness, within the compass of one person's senses, its connection between what he saw and what he heard, recorded faithfully. Muir creates an astounding visual image in his words, but it is the sound that seems to reverberate on the very page, that transmits the experience, first and foremost. Not only does this continue to exist in the memory, but also Muir takes us on his own imaginative journey at the end, suggesting that if this was caused by just *one* mountain reshaping itself, can we conceive what the sound of the creation might have been? In that sense, it is indeed truly Biblical, and fulfils the criteria we set ourselves when considering the work of John Clare, the sense of what it must have been like to be not only in the place, but also at that very moment in time. His images of the Sierra are wide angle location recording at its most opportunistic, vivid and accurate.

Yet equally, Muir has in his work a capacity to focus on the minute to which Emily Dickinson would relate. He devotes a page and a half to his celebration – there is no other word for it – of a small grasshopper. 'Wonderful,' he writes, 'that these sublime mountains are so loudly cheered by a creature so queer'.[13] Muir truly loves the grasshopper, is amused by it, and tunes his mental listening apparatus to its fragile sound:

Up and down a dozen times or so he danced and sang, then alighted to rest, then up and at it again. The curves he described in the air in diving and rattling ...

[12] Ibid., p. 462.
[13] Ibid., p. 222.

When he was on the ground he made not the slightest noise, nor when he was simply flying from place to place, but only when diving in curves, the motion seemed to be required for the sound.[14]

The account is full of the poignancy of an awareness of the tiny and frail creature set against the vastness of the terrain, as Muir records this drama within the context of the place itself, high on the mountainside:

A fine sermon the little fellow danced for me on the Dome, a likely place to look for sermons in stones, but not for grasshopper sermons. A large and imposing pulpit for so small a preacher. No danger of weakness in the knees of the world while Nature can spring such a rattle as this. Even the bear did not express for me the mountain's wild health and strength and happiness so tellingly as did this comical little hopper.[15]

In Muir's account of his childhood, *The Story of My Boyhood and Youth*, he describes his family's emigration from Dunbar to a farm at Portage, Wisconsin. There is an eloquent chapter in which he reveals the amazement of an eleven-year-old boy at the sounds and sights of birdlife he encountered there, noting that 'none of the bird people of Wisconsin welcomed us more heartily than the common robin … how we admired the beauty and fine manners of these graceful birds and their loud cheery song of *fear not, fear not, cheer up, cheer up*'.[16] Muir calls his chapter 'A Paradise of Birds'. It is a detailed account of sound and its effect on a young boy; here we can only dip into it and hint at its range, depth and precision of description. He notes, for example, that the thrush, or thrasher, 'makes haste to the topmost spray of an oak tree and sings loud and clear with delightful enthusiasm until sundown'. As other birds join in, 'their rich and varied strains make the air fairly quiver'.[17] Then comes a glowing passage on the bobolink, 'gushing, gurgling, inexhaustible fountains of song, pouring forth floods of sweet notes over the broad Fox River meadows in wonderful variety and volume'. Where Emily Dickinson hears a chorister, Muir finds himself absorbed in the outpourings of a full choir, every one of them a virtuoso soloist:

It seemed marvellous to us that birds so moderate in size could hold so much of this wonderful song stuff. Each one of them poured forth music enough for a whole flock, singing as if its whole body, feathers and all, were made up of music,

---

[14] Ibid.
[15] Ibid., p. 223.
[16] J. Muir, *The Story of My Boyhood*, p. 82.
[17] Ibid.

flowing, glowing, bubbling melody interpenetrated here and there with small scintillating prickles and spicules.[18]

When John Muir met his contemporary, John Burroughs (1837–1921) in July 1896, it was a joining of two minds united by nature, albeit coming to their subject from different aspects and directions. Burroughs, writing in his journal, declared that he found Muir 'a little prolix at times. You must not be in a hurry, or have any pressing duty, when you start his stream of talk and adventure … He is a poet and almost a Seer. Something ancient and far-away in the look of his eyes.'[19] Burroughs himself had been born a year before Muir and seven years after Emily Dickinson, on a farm at Roxbury in Delaware County, New York. He became well known in his lifetime as a naturalist and nature essayist, and an active member of the growing conservation movement in the United States. Ironically, during his lifetime, he was friends with a number of American captains of industry, whose working practices were to contribute to the climatic undermining of the planet. Among them, Henry Ford, (who gave him a car, one of the first in the Hudson Valley), Harvey Firestone and Thomas Edison, were all part of Burroughs's circle of friends. Yet with the ability to perceive, despite the nature of some of these acquaintanceships, the damage industrial progress would cause to the natural world, Burroughs was often vocal in his criticism of the price of industrial prosperity. His great hero was Ralph Waldo Emerson, and he was a close friend of Walt Whitman. In placing Burroughs in our story, we should see him as more of a literary naturalist than a scientific one. He worked, and was known, within a culture of the middle and upper classes of post-Civil War America where increasing mobility led to a new recreational experience of nature. Burroughs's biographer, Edward J. Renehan, explores this context:

> The same affluent Easterners who purchased books and read literary magazines such as *Scribner's* and the *Atlantic Monthly* also set great store in country houses, gardens, and annual vacations that included hiking … As the American industrial state expanded in the years after the Civil War, it created a new middle class with the wherewithal and leisure to pursue picturesque rural landscapes … Through his more than two dozen books of nature essays … John Burroughs provided a steady stream of encouragement, instruction, and inspiration for men and women of the educated classes who chose to take up hiking and

---

[18] Ibid., pp. 82–3.
[19] J. Burroughs, 'Journal', quoted in E. J. Renehan, *John Burroughs: An American Naturalist* (Post Mills, VE: Chelsea Green, 1992), p. 205.

nature study as an antidote to a society increasingly mortgaged to advance of technology and the rise of cities.[20]

In an 1877 book, *Birds and Poets*, he wrote, 'It might almost be said that the birds are all birds of the poets and of no one else, because it is only the poetical temperament that fully responds to them. So true is this, that all the great ornithologists – original namers and biographers of the birds – have been poets in deed if not word.'[21] Elsewhere, he wrote that 'I suspect it requires a special gift of grace to enable one to hear the bird-songs; some new power must be added to the ear, or some obstruction removed. There are not only scales upon our eyes so that we do not see; there are scales upon our ears so that we do not hear.'[22] If some of his pronouncements could sound somewhat patronizing, he also sought to qualify and justify them: 'Human and artificial sounds and objects thrust themselves upon us; they are within our sphere, so to speak; but the life of nature we must meet halfway; it is shy, withdrawn, and blends itself with a vast neutral background. We must be initiated; it is an order the secrets of which are well guarded.'[23]

Burroughs was a generous sharer of knowledge, and quick to acknowledge the work of others. When he was told that his ideas on nuthatches excavating their own nests disagreed with the findings of the ornithologist, Cordelia Stanwood (1865–1958), he is reported to have said: 'Well I know Miss Stanwood – she is a real bird woman, and if she has seen this trait, I believe absolutely what she says must be true.'[24] Stanwood, who lived her long life in the town of Ellsworth, Maine, had an ear for minute natural sounds akin to Emily Dickinson, and it was a music that brought her great comfort, and upon which she commented eloquently and often, as in her field notebook in May 1913: 'I lie under a big pine and listen to the black-throated green warbler that sings and feeds among its branches, to the purple finch that soars in an ecstasy of song. I feel rested, soothed, delighted with the world and myself.'[25] Notebooks such as this accompanied her throughout her life, and provided first-hand witnesses to her listening. Myers

---

[20] E. J. Renehan, *John Burroughs: An American Naturalist* (Post Mills, Vermont: Chelsea Green Publishing Company, 1992), p. 1.
[21] J. Burroughs, *Birds and Poets* (Frankfurt. Outlook, 2019), p. 4.
[22] J. Burroughs, in C. Z. Walker (ed.), *The Art of Seeing Things* (Syracuse: Syracuse University Press, 2001), p. 30.
[23] Ibid., p. 15.
[24] J. Burroughs, quoted by M. M Bonta. *Women in the Field: America's Pioneering Naturalists* (College Station: Texas A & M University Press, 1991), p. 217.
[25] C. Stanwood, Quoted by Myers Bonta, p. 212.

Bonta has extracted some examples, demonstrating that for Stanwood, birdsong brought more than aesthetic pleasure:

> Chickadees had a song 'like a paean of victory, sura-suree, sura-suree', and northern juncos 'a liquid trill that sounded like water gurgling over hidden pebbles'. A purple finch, she said, sings 'as if in ecstasy, as if it would be a joy to float and float and sing forever, but finally comes down quivering into the treetops', and 'the song of the winter wren is high, liquid, wild'.[26]

Stanwood's a gift for finding the world in the intensely local was one that John Burroughs understood instinctively. Roxbury had been his birthplace, and when circumstances enabled it, he returned and created an estate called 'Riverby' not far from his roots, where he lived and wrote for many years. He identified with the familiar: when he visited Selborne during a trip to England, he noted that the village had changed little since Gilbert White's time, observing that 'Selborne is as provincial as Roxbury'.[27] His friendships were fruitful, although not blind. He could see what America's industrialists were doing to the land, and was not uncritical of the racist elements in the thinking of his great friend, Walt Whitman. Equally, these same friends respected his knowledge of the natural world, and it informed their own work. After a discussion with Burroughs about the song of the hermit thrush, Whitman wrote in his notebook: 'Sings softer after sundown ... is very secluded ... his song is like a hymn ... never sings near farm houses – never in the settlement – is the bird of the solemn primal woods and of Nature pure and holy'.[28] At this time, Whitman was working on his elegy for Abraham Lincoln, and in his poem, 'When Lilacs last in the Dooryard Bloom'd' we find the sad song of the hermit thrush, preserved from that conversation:

> In the swamp in secluded recesses,
> A shy and hidden bird is warbling a song.
>
> Solitary the thrush,
> The hermit withdrawn to himself, avoiding the settlements,
> Sings by himself a song.
>
> Song of the bleeding throat,
> Death's outlet song of life (& c.)[29]

---

[26] Ibid., pp. 215–16.
[27] Quoted by Renehan, p. 147.
[28] Ibid., p. 83.
[29] W. Whitman, 'When Lilacs Last in the Dooryard Bloom'd' in *Leaves of Grass* (New York: Modern Library, 2001), p. 409.

Burroughs knew the bird intimately, had listened intently to its unique flute-like song, and had thought long and hard about the nature of it, considering that not even one of the great fathers of American ornithology, John James Audubon (1785–1851), had, in his view, quite succeeded in capturing its elusive sound in his reflections. Burroughs as an infant had arrived in the Audubon era; the great *Birds of America* project was nearing completion when he was born, and his early field trips during the Civil War had been informed by a fascination drawn from his viewing of a portfolio edition of the vast book in the library of the Military Academy at West Point. Today it is largely through Audubon's stunning illustrations of birdlife, which reached their zenith in *The Birds of America*, that we remember his work; yet his writings too hold much interest, as they did for Burroughs. Audubon's ear was as shrewd as his eye, and his quest for precision in his description of sound was exact to the point of bordering on the eccentric. For example, here is his capture in words of the little screech owl: 'The notes of this owl are uttered in a tremulous, doleful manner, and somewhat resemble the chattering of the teeth of a person under the influence of extreme cold, although much louder. They are heard at a distance of several yards and by some people are thought to be of ominous import'.[30] His description of the barred owl too, which is included in *The Birds of America* on plate 46, drew a human comparison in its textual description of its sound:

> Were you ... to visit that happy country [the state of Louisiana], your ear would suddenly be struck by the discordant screams of the barred owl. Its *whah, whah, whah, whah-aa* is uttered loudly and in so strange and ludicrous a manner that I should not surprise you, kind reader, when you and I meet, to compare these sounds to the affected bursts of laughter which you may have heard from some of the fashionable members of our own species.[31]

The analogy offers a sound image as precise as it is satirical, ironic and amusing. Audubon clearly had an ear for audio, familiar as he was with music and language, explanations for this to become apparent through a brief examination of his biography. He was born in the French colony of Saint-Domingue (now Haiti) on his father's sugarcane plantation. After a period in France during the turbulent years of revolution, in 1803 the eighteen-year-old Audubon emigrated to America using a false passport obtained by his father to enable him to

---

[30] J. J. Audubon, in E. Rhodes (ed.) *The Audobon Reader* (New York: Alfred A. Knopf, 2006), p. 149.
[31] Ibid., p. 139.

escape conscription in the Napoleonic Wars, and thereafter, through numerous adventures and side-tracks, the young man's interest in the natural world burgeoned. His written accounts developed in parallel with his art, and there is a clarity and poetry combined in his writing that creates another, complementary kind of pictorial image in the mind. To view the plates in *The Birds of America* is only part of understanding his relationship to the experience of nature; he would often write short essays placing the birds under his consideration into a context, as with his text on the chuck-will's-widow, an American member of the nightjar family, found in the south eastern states near swamps, rocky uplands and pine woods, which is portrayed in plate 52 of *The Birds of America*. Also, as with the English counterpart, carrying the name 'goatsucker', Audubon found the strange sound to be restful and haunting, and it is worth comparing his textual evocation with that of Gilbert White and John Clare in previous chapters:

> About the middle of March the forests of Louisiana are heard to echo with the well-known notes of this interesting bird. No sooner has the sun disappeared and the nocturnal insects emerge from their burrows than the sounds 'chuck-will's widow', repeated with great clearness and power six or seven times in as many seconds, strike the ear of every individual, bringing to the mind a pleasure mingled with a certain degree of melancholy which I have often found very soothing.[32]

In all cases, there is a sense, not only of the sound itself, but also of the place in which the sound occurs. Burroughs had written that 'the song is not all in the singing, any more than the wit is all in the saying. It is in the occasion, the surroundings, the spirit of which it is the expression.'[33] It is after all, in a specific environment that the wild track is to be found, the sound of the place being the continuing and surrounding ambience. All the writers explored in this chapter were able to evoke intensely the auditory presence of their particular landscapes, however, varied those landscapes were in terms of scale and topography. Giving in to a migratory nostalgia, Burroughs wrote that 'nearly every May I am seized with an impulse to go back to the scenes of my youth, and hear the bobolinks in the home meadows once more. I am sure they sing better than anywhere else.'[34] Place and time: the changing music of the seasons that Emily Dickinson had

---

[32] Ibid., p. 143.
[33] J. Burroughs, *The Art of Seeing Things*, p. 33.
[34] Ibid.

noted with a mixture of pain and wonder, that had given Cordelia Stanwood such peace of mind, all within the context of an intimately known local world.

It was this mix that fuelled the writing of Henry David Thoreau, (1817–62) who kept a journal that gradually refined itself towards a project he was unable to complete, his *Kalendar*, that would have focused, as everything else did in his life, on the terrain of Concord, Massachusetts. Thoreau belonged to what became known as 'The Concord School' a group of writers and poets who gathered around the essayist and philosopher Ralph Waldo Emerson (1803–82) during the 1830s. In 1836, two years after he had moved to Concord, Emerson had published an influential essay entitled 'Nature', in which he formulated and expressed the philosophy of transcendentalism, which views the phenomenological world as a form of symbol of the inner life. He had met and talked with Wordsworth and Coleridge in England, and his thinking influenced a whole generation of writers and poets. Thoreau himself was born in Concord, and lived there for much of his life. For a time he worked as a handyman for Emerson, at the same time absorbing his philosophy, while growing in his own fascination with the natural world around him. When he was twenty he began to keep a detailed notebook, containing quotes, poems short essays and observations, which gradually evolved into what has become known as his journal, although it was a diary in a less than formal sense, and more a source book of responses. As the editor of the *Journal, 1837–1861* points out, it is not 'literally what Thoreau wrote each day; he often wrote up entries days later, from notes, and … would also go back years later and make further additions and corrections … the Journal is a record of what he and Nature did on a given day, and how those doings affected each other'.[35] The sounds he noted, frequently acknowledge the world around him, and the poet in him would seek out a simile or metaphor to explain the exact phenomenon of what he heard. In an entry for August 1851, he set out his priorities as a writer: 'I omit the unusual – the hurricanes and earthquakes – and describe the common. This has the greatest charm and is the true theme of poetry. You may have the extraordinary for your province if you will let me have the ordinary. Give me the obscure life'.[36] It is in his observation of minutiae that he shows himself truest to this principle. For example, this entry, from 3 May 1852, written when he was thirty-four.

---

[35] D. Searls, in H. D. Thoreau (ed.) *The Journal, 1837–1861* (New York: New York Review of Books Publishing, 2009), p. xiv.
[36] Thoreau, ibid., p. 71.

> Going through the Depot Field, I heard the dream frog at a distance. The little piping frogs make a background of sound in the horizon, which you do not hear unless you attend. The former is a trembling note, some higher, some lower, along the edge of the earth, an all-pervading sound. Nearer, it is a blubbering or rather bubbling sound, such as children, who stand nearer to nature, can and do often make.[37]

And two days later: 'Heard the first cricket singing, on a lower level than any bird, observing a lower tone – the sane, wise one – than all the singers. He came not from the south, but from the depths. He has felt the heat at last, that migrates downwards.'[38] The tiny creature held an enduring fascination for him. On 4 August 1851, there is this: 'I heard the note of an autumnal cricket, and was penetrated with the sense of autumn. Was it sound? or was it form? or was it scent? or was it flavour?'[39] These tiny sounds were what drew him, time after time: 'Why was there never a poem on the cricket?' he writes on 3 September: 'Its creak seems to me to be one of the most prominent and obvious facts in the world, and the least heeded.'[40] That same day, he noted the humming of the telegraph wires, a new scientific and communication phenomenon that would transform the world. He became fascinated by the sound, both for the idea that it was a symptom of what the wires carried, but equally for the way their presence interacted with the natural world through which they spread: 'As I went under the new telegraph-wire, I heard it vibrating like a harp high overhead. It was the sound of a far-off glorious life, a supernal life, which came down to us, and vibrated the lattice-work of this life of ours.'[41] Of Thoreau's thinking, Emerson wrote that 'his eye was open to beauty, and his ear to music. He found these, not in rare conditions, but wheresoever he went. He thought the best of music was in single strains; and he found poetic suggestion in the humming of the telegraph-wire.'[42] When it came to birdsong, he felt the effort of words, or even human music, would always be inadequate:

> When I hear a bird singing, I cannot think of any words that will imitate it. What word can stand in place of a bird's note? You would have to surround it with a *chevaux de frise* of accents, and exhaust the art of the musical composer

---

[37] Ibid., p. 133.
[38] Ibid.
[39] Ibid., p. 65.
[40] Ibid., p. 73.
[41] Ibid.
[42] R. W. Emerson, 'Thoreau' in *Nature and Selected Essays* (London: Penguin Classics, 2003), p. 408.

besides with your different bars, to represent it, and finally get a bird to sing it, to perform it. It has so little relation to words.[43]

Yet for Thoreau, the written word is also 'the choicest of relics', as he was to write in his masterpiece, *Walden*; '[It is] the work of art nearest to life itself. It may be translated into every language, and not only read but also actually breathed from all human lips; – not be represented on canvas or in marble only but be carved out of the breath of life itself'.[44] As for the Journal, it spans almost a lifetime, and overarches – while complementing – his other work. On 4 July 1845, Thoreau moved to a hut he had constructed on Emerson's land at Walden Pond, where he remained for two years, two months and two days, until 6 September 1847, a sojourn from which grew *Walden*, published in 1854. Thoreau noted that 'this small lake was of most value as a neighbour in the intervals of a gentle rainstorm in August, when, both air and water being perfectly still, but the sky overcast, mid-afternoon had all the serenity of evening, and the wood-thrush sang around, and was heard from shore to shore.'[45] At one point, Thoreau devotes a chapter specifically to the sounds surrounding him; we hear with him, the voices of nature as they flood back into the consciousness as the senses calm. One Sunday, he becomes aware of church bells, their sound coming from various sources in surrounding communities: Lincoln, Acton, Bedford or Concord itself. He notices that, when the breeze is propitious, these sounds become part of the natural world; this is not intrusion, but a blending, like the humming of the telegraph wires. A bell, after all, fades through the life of a single sounding towards silence, but the moment of inaudibility, where it melds with the ambient sounds around it, is subtle and elusive. Thoreau captures the idea perfectly:

> At a sufficient distance over the woods this sound acquires a certain vibratory hum, as if the pine needles in the horizon were the strings of a harp which it swept. All sound heard at the greatest possible distance produces one and the same effect, a vibration of the universal lyre, just as the intervening atmosphere makes a distant ridge of earth interesting to our eyes by the azure tint it imparts to it. There came to me in this case a melody which the air had strained, and which had conversed with every leaf and needle of the wood, that portion of the sound which the elements had taken up and modulated and echoed from vale to vale. The echo is, to some extent, an original sound, and therein is the magic and

---

[43] H. D. Thoreau, *Journal*, p. 137.
[44] H. D. Thoreau, *Walden* (London: Penguin Books, 2016), p. 96.
[45] Ibid., p. 81.

charm of it. It is not merely a repetition of what was worth repeating in the bell, but partly the voice of the wood.[46]

There is an extraordinary sensitivity to Thoreau's sonic tuning; he gave the best possible council to both sound recordist and poet alike when he wrote 'no method nor discipline can supersede the necessity of being forever on the alert'.[47] In other words, remind yourself when you are *not* seeing and listening. His account of screech owls compares the sound to 'mourning women ... their dismal scream is truly Ben Jonsonian. Wise midnight hags!' He seeks a precision, the exact words for it, not the 'honest and blunt tu-whit tu-who of the poets', no, not that old cliché, but 'the mutual consolations of suicide lovers remembering the pangs and delights of supernal love in the infernal groves'.[48] It is a sound he loves, this strange cross-over between supernatural music and birdsong, all along the wood-side, almost as though it could be the woods and the lake itself issuing forth the sounds. It is a dark underside of human music, 'the low spirits and melancholy forebodings of fallen souls that once in human shape night walked the earth and did the deeds of darkness, now expatiating their sins with their wailing hymns'. Almost trance-like, he listens, finally translating the sound into human language: 'They give me a new sense of the variety and capacity of that nature which is our common dwelling'.

> *Oh-o-o-o-o that ever I had been bor-r-r-r-n!* sighs one on his side of the pond, and circles with the restlessness of despair on some new perch on the grey oaks. Then *– that ever I had been bor r r r n!* echoes another on the farther side with tremulous sincerity, and *– bor-r-r-r-n!* comes faintly from far in the Lincoln woods.[49]

In the work of Thoreau, we come again and again, both in *Walden* and his *Journal*, to a sense of a place as the context in which the specifics of sound are heard and noted: 'As I sit at my window this summer afternoon, hawks are circling about my clearing; the tantivy of wild pigeons, flying in twos and threes athwart my view, or perching restless on the white-pine boughs behind my house, gives a voice to the air.'[50] He is often about the natural detail, but that detail is a symptom for him of something beneath and beyond the physical world: invisible, like

---

[46] Ibid. p. 115.
[47] Ibid. p. 104.
[48] Ibid., pp. 116–17.
[49] Ibid. p. 117.
[50] Ibid. p. 107.

sound. He loves an intimate world that has visual boundaries, beyond which other sounds filter in, and upon which he can meditate. Emerson wrote of Thoreau: 'His power of observation seemed to indicate additional senses. He saw as with microscope, heard as with ear-trumpet, and his memory was a photographic register of all he saw and heard. And yet none knew better than he that it is not the fact that imports, but the impression or effect of the fact on your mind.'[51] It was as though the senses became a portal through which the physical world revealed another plain of existence, something running in parallel with the phenomenological world. Alongside this bubble of metaphysical feeling, there was the ever-growing sound of the new industries, the scientific studies that would apparently explain everything. And yet ... Thoreau sometimes seemed to move seamlessly between the physical and the etheric within a single experience, a state brought on by an intense 'oneness' with a particular place. Near the end of his essay, *Walking*, there is this: 'I took a walk on Spaulding's Farm the other afternoon'. For a few lines, he describes the scene, a peaceful, golden view, lit by the declining sun over the pine woods, the glow filtering through the wood 'as into some noble hall.'[52] Then things begin to change; he starts to gain the impression that 'some ancient and altogether admirable and shining family had settled there in that part of the land called Concord, unknown to me – to whom the sun was a servant – who had not gone into the society of the village – who had not been called on.'[53] This unseen but sensed family belong to the place, yet they neither know, nor are they known by, the farmers and other inhabitants of the locality. It continues, increasingly dream-like, with an absolute certainty of presence within the woods, as the voices of nature absorb and envelope his whole being:

> Their attics were the tops of the trees. They are of no politics. There was no noise of labour. I did not perceive that they were weaving or spinning. Yet I did detect, when the wind lulled and hearing was done away, the finest imaginable sweet musical hum, as of a distant hive in May, which perchance was the sound of their thinking.[54]

Dickinson, Muir, Burroughs, Stanwood and Thoreau provide us with vivid witnesses to the voice of the natural world, and belong to the evolving expression of

---

[51] Emerson, *Nature and Selected Essays* (London: Penguin Classics, 2003), p. 406.
[52] H. D. Thoreau, *Walking* (Los Angeles: Enhanced Media Publishing, 2017), p. 48.
[53] Ibid.
[54] Ibid.

the sound-based written texts of nature, whether in their expression of minutiae, the mystery of murmuring twilight, or in the sublime epic scale of epic landscapes and distance. As for Thoreau himself: in 1860, he read Charles Darwin's *Origin of Species*. Later that same year, he caught a cold, which developed into bronchitis. He had contracted tuberculosis years earlier, in 1835, and had suffered from it sporadically thereafter. Now, his condition worsened rapidly, but he spent his last month's editing his work, including his journal and his essays. Around him, the hum of the wires grew louder as the world changed; the Civil War raged, and in London's South Kensington, the International Exhibition opened, boasting amongst its exhibits, the developments and promises of science. On 6 May 1862, he died, aged just forty-four. Asked by a relative shortly before the end, if he had made his peace with God, he is famously said to have replied: 'I was not aware we had ever quarrelled.'[55]

---

[55] H. D. Thoreau, *Journal*, p. xxiv.

10

Survival and the sound of spirit

Sound is invisible; it moves through time, as do we. It emerges or strikes, it fades or ceases, leaving silence. It is often ghostly, unearthly until it is explained or given context; it can belong to what Sigmund Freud called 'das unheimliche' – the uncanny. In September or October, 1832, Charles Darwin, then staying at Bahia Blanca in Argentina, heard something he did not understand. It was 'a singular, deep-toned, hissing note: when I first heard it, I thought it was made by some wild beast, for it is a sound that one cannot tell whence it comes, or from how far distant'.[1] It turned out to be a male ostrich, and Darwin happened to have encountered it during the nesting season. The sense of threat implied in the sound may well have been very real, because 'it is asserted that at such times they are occasionally fierce, and even dangerous, and that they have been known to attack a man on horseback'.[2] In order to gain an idea of their original cultural impact, we should attempt to read the works of Darwin, and his near contemporary, Alfred Wallace (1823–1913) as they were encountered, heard and responded to when they were first published, in a world where discovery was both scientific and imaginative. Alfred Russel Wallace was, with Darwin, the co-discoverer of the theory of evolution by natural selection, was also the pre-eminent tropical botanist of his day, and has come to be regarded as the founder of evolutionary biogeography. His account of his travels between 1854 and 1862 across South-East Asia, from Singapore to western New Guinea was published in 1869 under the title, *The Malay Archipelago*.

'Returning to Ampanam', Wallace writes, 'I devoted myself for some days to shooting the birds of the neighbourhood.'[3] Not, however, before he had committed to memory and then to paper, the sound of the Tropidorhynchus timoriensis, ('allied to the Friar bird of Australia',) which were 'here called "Quaich-quaich", from their strange loud voice, which seems to repeat these

---

[1] C. Darwin, *The Voyage of the Beagle* (New York: Meridian Books, 1996), pp. 77–8.
[2] Ibid.
[3] A. R. Wallace, *The Malay Archipelago* (London: Penguin Classics, 2014), p. 172.

words in various and not unmelodious intonations'.[4] Wallace's writings often evoke a vivid sense of place, and he shows himself to be a sensitive recorder of experience as a context for his studies:

> The country around was pretty and novel to me, consisting of abrupt volcanic hills enclosing flat valleys or open plains. The hills were covered with dense scrubby bush of bamboos and prickly trees and shrubs, the plains were adorned with hundreds of noble palm-trees, and in many places with a luxuriant shrubby vegetation. Birds were plentiful and very interesting ... Small white cockatoos were abundant, and their loud screams, conspicuous white colour, and pretty yellow crests, rendered them a very important part of the landscape.[5]

The complementary sound contribution of the fauna to the visual impressions of the flora are often documented in Wallace's writing. Of the Great Bird of Paradise, he writes that while 'the full-plumaged birds are less plentiful, their loud cries, which are heard daily, show they are ... very numerous. Their note is "Wawk-wawk-wawk – Wōk, wōk-wōk", and is so loud and shrill as to be heard at a great distance, and to form the most prominent and characteristic sound in the Aru Islands'.[6] Sound is part of the landscape, creating both an impression and a memory capable of evoking a sense of the place long after it has been left behind. Wallace and Darwin were part of a generation of naturalists, scientists and explorers who became familiar with the thoughts and findings of the German physicist and physician, Hermann von Helmoltz (1821–94), known for his ideas on the perception of not only space and colour, but also on the sensation of tone, and of sympathetic resonance, which explored the idea of the ear as an instrument capable of responding to musical notes as vibrations. Darwin drew directly on this thinking in his book, *The Descent of Man*. 'With respect to sounds', Darwin writes:

> Helmholtz has explained to a certain extent on physiological principles, why harmonies and certain cadences are agreeable ... Whether we can or not give any reason for the pleasure thus derived from vision and hearing, yet man and many of the lower animals are alike pleased by the same colours, graceful shading and forms, and the same sounds.[7]

---

[4] Ibid.
[5] Ibid., p. 173.
[6] Ibid., p. 588.
[7] C. Darwin, *The Descent of Man* (London: Penguin Classics, 2004), p. 115.

At the same time, many religious commentators of the nineteenth century joined the debate into the nature of sound and its appeal to the mind. As E. J. Gillin has commented in his book, *Sound Authorities*, 'these commentators endeavoured to define the physical nature of sound and explore why music was aesthetically pleasing'.[8] Among them was W. Weldon Champneys, Dean of Lichfield, who gave a sermon in 1872 in which he explained how sound travelled to the brain via the ears through the air, while declaring the whole to be divinely created, including the aesthetics of musical appreciation itself: 'Wonderful it is that the air which is our life should also be the instrument of our pleasure. If there was no air, there would be neither life nor music'.[9] Darwin on the other hand had argued that music, in the form of modulated sound in support of species propagation, had grown initially, 'as a separate evolutionary function of sexual selection, preceding language'.[10] From either standpoint, a key factor concerned the mysteries of the invisible, that which remained out of sight but heard, including an unexplained sound. This in turn generated a dialogue which drew further attention to the presence of natural phenomena in juxtaposition with growing industrial human noise. Into this came transcendentalism and a resurgence of interest in Pantheism that was the viewpoint of many leading writers and philosophers of the nineteenth century, attracting poets such as Wordsworth, Coleridge and Whitman, and Emerson and Thoreau in the United States. So prevalent did it become that the Vatican found it necessary to intervene on the subject and issue a condemnation by Pope Pius IX in the *Syllabus of Errors* in 1864.[11] Yet Pantheism, the idea that identifies the universe as a manifestation of God or gods, dates back thousands of years. The word was coined in 1697 by the mathematician Joseph Raphson, but the concept existed long before the term. The invisibility of sound, combined with its ubiquity, has been frequently interpreted as a symptom of something supernatural, an unseen world underlying our existence. For many theologians, the idea of a god *within* things was a complex argument, while for others, such as the Jesuit poet Gerard Manley Hopkins, writing in his journal for 3 June 1866, while a 22-year-old

---

[8] E. J. Gillin, *Sound Authorities: Scientific and Musical Knowledge in Nineteenth-Century Britain* (Chicago: University of Chicago Press, 2022), p. 203.
[9] W. W. Champneys, quoted ibid., p. 223.
[10] E. J. Gillin, *Sound Authorities*, p. 223.
[11] *The Syllabus of Errors* condemns a total of eighty heresies, articulating Catholic Church teaching across a range of philosophical and political questions. It remains controversial.

student at Oxford, everything seemed to be present within an intense observation of nature:

> The green was softening to grey. The meadows yellow with buttercups and under-reddened with sorrel and containing white oxeyes and puffballs. The cuckoo singing one side, on the other from the ground and unseen the woodlark, as I suppose, most sweetly with a song of which the structure is more definite than the skylark's and gives the link with that of the rest of birds.[12]

Hopkins would return to contemplation of both birds in his poetry; in 1876, his poem, 'The Woodlark' begins with a continuation of a quest for the source of the bird's song:

> Teevo cheevo cheevio chee:
> O where, what can that be?
> Weedio-weedio: there again!
> So tiny a trickle of song-strain
> And all round not to be found
> For brier, bough, furrow, or green ground
> Before or behind or far or at hand
> Either left either right
> Anywhere in the sunlight (& c.)[13]

The woodlark is invisible to him, so the sound is its only manifestation. It is as though it becomes the very voice and articulation of the natural world. Hopkins, meanwhile, makes a silent sound out of words. Yet beyond all this page-music, there is the communication of a oneness between nature and poet that reaches a kind of religious ecstasy. The skylark too was captured in verse in a sonnet by Hopkins, drafted first in 1877, and initially called 'Walking by the Sea', but later retitled and reworked as 'The Sea and the Skylark'. It expresses a sense of the landscape – or rather here, the seascape – as a single ancient music born of a mutual creation:

> On ear and ear two noises too old to end
>   Trench – right, the tide that ramps against the shore;
>   With a flood or a fall, low lull-off or all roar,

---

[12] G. M. Hopkins (ed.), H. House and G. Storey, *The Journals and Papers of Gerard Manley Hopkins* (London: Oxford University Press, 1959), p. 138.
[13] G. M. Hopkins (ed.) and N. H. Mackenzie, *The Poetical Works of Gerard Manley Hopkins* (Oxford: Clarendon Press, 1990), p. 131.

> Frequenting there while moon shall wear and wend.
> Left hand, off land, I hear the lark ascend,
> > His rash-fresh re-winded new-skeinèd score
> > In crisps of curl off wild winch whirl, and pour
> > And pelt music, till none's to spill nor spend.[14]

Thus, the octave of the poem; for the rest, these sounds shame the 'shallow and frail town' and indeed our own 'sordid turbid time, /Being pure!' But as with so much in Hopkins's work, the sound of the words, the forward-driving sprung rhythm of the verse, makes its own heady music. This stanza is pure stereo – even surround-sound – in its effect. Here is the music of the natural world, pulled from left and right to form a whole.

Many of these poems are difficult to date, and Hopkins frequently reworked verses over several years. In an early journal entry, he refers to the cuckoo with a stereo presence, a strong sense of direction and perspective in his listening, and he comes back to the sound of the bird often, as here, from an entry from June 1873: 'Sometimes I hear the cuckoo with wonderful clear and plump and fluty notes: it is when the hollow of rising ground conceives them and palms them up and throws them out, like blowing into a big humming hewer'.[15] This found its way into a short poem which surely must have grown out of pure auditory experience, blended with note-taking from the journals and elsewhere:

> Repeat that, repeat,
> Cuckoo, bird, and open ear wells, heart springs, delightfully sweet,
> With a ballad, with a ballad, a rebound

Off trundled timber and scoops of the hillside ground, hollow hollow hollow ground:

> The whole landscape flushes on a sudden at a sound[16]

As to the location of this poem, Hopkins writes in the same journal entry, of the sound heard 'another time from Hodder wood when we walked on the other side of the river'.[17] An April entry that same year[18] has him staying at Whitewell, a village in the Forest of Bowland, at the time part of Yorkshire, and

---

[14] Ibid., p. 143.
[15] G. M. Hopkins, *Journal*, p. 232.
[16] G. M. Hopkins, *Poems*, p. 158.
[17] G. M. Hopkins, *Journal*, p. 232.
[18] Ibid., p. 230.

standing above a bend of the River Hodder. This then, we may believe, is how the soundscape of a moment manifested itself on a late spring day in 1873 in a north country wood.

Hopkins trained as a Jesuit, but his great spiritual influence came through the teachings of Duns Scotus (c.1265–1308), a Scottish Catholic priest and Franciscan friar, whose best-known doctrine became known as the univocity of being, propagating the view that existence is the most abstract concept possible, and can be applied to everything that is, and he extended this to the existence of God. Inevitably, this would prove controversial, particularly within some more conservative channels of Catholicism. Hopkins himself was able to reconcile his religious calling with his poetic vocation through his reading of Duns Scotus, particularly in the nature of 'haecceity', which Scotus translates from the Latin, 'haecceitas' as 'Thisness'. Hopkins felt that everything in the universe could be characterized by what he called 'inscape', a distinctive design at the root of individual identity. The power to recognize this leads to an understanding of an object or creature as being subject to an intense thrust of energy, a force which he called 'instress'. Sound is a link between this and the material, and Hopkins expresses the idea vividly in the soundscape of his sonnet on the kingfisher; once again, the octave pouring its meaning through sound and light:

> As kingfishers catch fire, dragonflies draw flame;
>   As tumbled over rim in roundy wells
>   Stones ring; like each tucked string tells, each hung bell's
> Bow swung finds tongue to fling out broad its name;
> Each mortal thing does one thing and the same:
>   Deals out that being indoors each one dwells;
>   Selves – goes its self; *myself* it speaks and spells,
> Crying *What I do is me: for that I came.*[19]

Sounds come and go, overlap, flow together, blend and shape themselves into being which for Hopkins, is the manifest form of God, 'for Christ plays in ten thousand places'. Indeed, in one of his most famous poems, 'Spring' he hears the thrush in the context of the place to the extent that the acoustic of one celebrates the initiating sound of the other like a church interior; it is a reference that Dafydd ap Gwyllim would have acknowledged:

---

[19] G. M. Hopkins, *Poems*, p. 141.

> thrush
> Through the echoing timber does so rinse and wring
> The ear, it strikes like lightnings to hear him sing.[20]

(Note that here it is the ear that is 'rinsed', and 'wring' is as of something from which sound is wrung out, like a piece of cloth. The song of the bird within the cathedral woodland cleanses and purifies the faculty of true hearing and listening). It is sound heard, and sound conveyed through the auditory imagination. He himself might well have expressed it thus:

> Everything gives forth a sound of life. The twittering of swallows from above, the song of greenfinches in the trees, the rustle of hawthorn sprays moving under the weight of tiny creatures, the buzz upon the breeze; the very flutter of the butterflies' wings, noiseless as it is, and the wavy movement of the heated air across the field cause a sense of motion and of music.[21]

In fact, those words were written not by Gerard Manley Hopkins, but by his almost exact contemporary, Richard Jefferies (1848–87), and they capture the poetic intensity of a school of nature writing that sought to grasp something beyond the science and the fact of cause and effect, reaching into and beyond consciousness itself.

Richard Jefferies was born two years after Hopkins, and he died at the age of thirty-eight, of tuberculosis, two years before him. His early years were spent on a farm in the village of Coate in Wiltshire, and produced a classic book about childhood, *Bevis*. Much of his short working life, however, was as a journalist, writing often for magazines and journals for the consumption of urban and suburban readerships. Although he wrote some novels such as his apocalyptic-pioneering work, *After London* and a full-length spiritual biography, *The Story of My Heart*, he was essentially a miniaturist specializing in a highly focused style of poetic, often ecstatic prose. For some he was a mystic, for others it was in his ability to convey with great precision his awareness of the natural world, and the people within the rural landscape, that marked him out as a writer of vivid imaginative powers. He could be over sentimental at times, and certainly his was a subjective style. Undeniably, however, he possessed an acute ability to be a 'receiver' of sound, with a remarkable skill to draw inner meaning from his senses. He was also capable of touching on the mystery of natural sound,

---

[20] Ibid., p. 142.
[21] R. Jefferies, *Nature near London* (London: Chatto & Windus, 1908), p. 6.

felt by Darwin and Wallace, a haunting that can (almost) always be explained, but which grows from the strangeness of new experience and has its answer in discovery:

> The loudest sound in the wood was the humming in the trees; there was no wind, no sunshine; a summer's day, still and shadowy, under large clouds high up. To this low humming the sense of hearing soon became accustomed, and it served but to render the silence deeper. In time, as I sat waiting and listening, there came the faintest far-off song of a bird away in the trees; the merest thin upstroke of sound, slight in structure, the echo of the strong spring singing (& c.)[22]

Jefferies often comes back to the ambient sound of nature, and then goes beyond it into what seems to be Hopkins's 'inscape'. He performs with his senses what we today would ask of a parabolic microphone, creating a field recording of a place in the very act of being. These observations sometimes turn into lengthy meditations, as in his essay, 'The Pageant of Summer', where once again, he finds himself deep in woodland. As his listening intensifies and layers under layers of sound begin to come to the surface.

> I become aware of a sound in the very air. It is not the midsummer hum which will soon be heard over the heated hay in the valley and over the cooler hills alike. It is not enough to be called a hum, and does but tremble at the extreme edge of hearing. If the branches wave and rustle they overbear it; the buzz of a passing bee is so much louder it overcomes all of it that is in the whole field.[23]

Jefferies is hearing a sound beneath sound; it is that mysterious 'noiseless noise in the orchard', to which Emily Dickinson referred in her letter to Thomas Wentworth Higginson, quoted in the last chapter.[24] The only way this 'almost sound' can be defined is by seasonal comparison; winter in this place, thinks Jefferies, would be silent. Just the creak and crack of branches and leaves and frost under foot, the air soundless. What we are hearing here, he tells us, is an inner sound, perhaps the very sound of life itself stirring:

> The sap moves in the trees, the pollen is pushed out from grass and flower, and yet again these acres and acres of leaves and square miles of grass blades ... are drawing their strength from the atmosphere. Exceedingly minute as these vibrations must be, their numbers perhaps may give them a volume almost

---

[22] R. Jefferies, *The Open Air* (London: Lutterworth Press, 1948), p. 92.
[23] R. Jefferies, *The Life of the Fields* (London. Lutterworth Press, 1947), p. 65.
[24] E. Dickinson, *Emily Dickinson: Letters*, p. 172.

reaching in the aggregate to the power of the ear. Besides the quivering leaf, the swinging grass, the fluttering bird's wing, and the thousand oval membranes which innumerable insects whirl about, a faint resonance seems to come from the very earth itself. The fervour of the sunbeams descending in a tidal flood rings on the strung harp of earth. It is this exquisite undertone, heard and yet unheard, which brings the mind into sweet accordance with the wonderful instrument of nature.[25]

The difficulty for someone coming fresh to the work of Richard Jefferies is its sheer diversity. He was a journalist who wrote of all aspects of country life: the daily life of labourers, the work of gamekeepers, village architecture and the social life of dwellers in the country. At the same time he was capable of transcendent mystical poetic prose such as this. Richard Mabey has written that 'it is still hard to say exactly what kind of writer Richard Jefferies was ... He can appear, sometimes inside a single piece of writing, as a small-town journalist, a romantic radical, a social historian and an apologist for the landowning class.'[26] The poet Jeremy Hooker has picked up the point, and offers a solution in the powers of objective observation that Jefferies possessed, which demonstrate his ability as an outsider to concentrate the mind and senses on a situation or place:

> The problem for the critic or anthologist of Jefferies is, of course, that the journalist who contributed articles to journals such as *The Live Stock Journal and Fancier's Gazette* is the same man whose writing is often discussed and anthologised in books about mysticism ... It is, I believe, in looking at Jefferies' art of seeing that we can perceive the one man whose diverse writings only appear to fragment.[27]

This 'art of seeing' is also the art of hearing, of listening beyond hearing, whole-mind attention to the sonic symptoms of the world. Jefferies lived a short life; he moved away from his home village, (although he remained spiritually 'there,') lived in the suburbs of London – in Sydenham and Eltham – for a time and died at Goring-by-Sea in Sussex. As his health declined, the poetry in his writing grew more precise in its perception of the sensory effect of being in a place. In his last essays, notably in works such as 'Hours of Spring', he mourns the

---

[25] Ibid., p. 66.
[26] R. Mabey, in R. Jefferies (ed.) *Landscape with Figures: An Anthology of Richard Jefferies's Prose* (London: Penguin Books, 1983), p. 7.
[27] J. Hooker, 'Richard Jefferies: The Art of Seeing' In *Writers in a Landscape* (Cardiff: University of Wales Press, 1996), p. 21.

fact that his health separates him from the natural world, while at the same time, all this will continue when he is gone, just as Emily Dickinson's fly continued to buzz after her imagined death: 'Today I have to listen to the lark's song – not out of doors with him, but through the window-pane, and the bullfinch carries the rootlet fibre to his nest without me'.[28] It is the fate of the outsider, the objective observer who comes to feel himself a member of a tribe that in the end has a culture and existence existing independently of him or her. Darwin and Wallace too, were by nature of their travels, discoveries and findings, outsiders. Jefferies, through being the kind of observer for whom the farm labourer and the minutest insect were each in turn worthy of his intensive scrutiny, came in the end to the realization of a oneness within nature that the fact of mortality condemned him to leaving behind. Yet it was through this quality of the outsider that all these writers found themselves capable of expressing their powers of seeing, listening and exploring in terms of humankind. It is part of the job, to be an outsider, just as it is for the poet, in order to find new ways of seeing, hearing and saying.

Jefferies found himself through health and circumstance, an outsider in a familiar world. Darwin and Wallace and others, for whom increased rather than reduced mobility brought a sense of objectivity born of being strangers in a foreign land, found a value in being culturally outside the world they explored. Travellers brought new perspectives to the study of natural history and the sound of terrains previously little known and certainly unheard, through their writings. Within that context, few could claim to be more of an outsider than W. H. Hudson (1841–1922) when it came to the intimate study of the English countryside. Born seven years before Jefferies, in Argentina, he came from a background with little education, and worked as a gaucho until his emigration to England in 1874, at the age of thirty-three. His great desire was to hear the sound of British birds, and his long life saw him become both an authority on the avian world and a champion for its protection and conservation; it was largely his efforts to stop the abuse of birds, that led to the foundation of the Royal Society for the Protection of Birds (RSPBs). Today an oil painting of Hudson, binoculars in hand, hangs in the RSPB's headquarters at Sandy in Bedfordshire. Thus his was a reversal of exploration as understood by Darwin and Wallace, coming to England with a similar sense of quest that others took with them to distant lands.

---

[28] R. Jefferies, 'Hours of Spring' in *Field and Hedgerow*, S. J. Looker (ed.) (London: Lutterworth Press, 1948), p. 23.

Like Jefferies, Hudson lived with mortality, having suffered when young with rheumatic fever, which weakened his heart. For a while at first, he lived a homeless life in his adopted land, and in London's Hyde Park, where he once slept, there is now a monument to him by Jacob Epstein. As he established himself, he took up residence from 1876 in St Luke's Road, Bayswater, London, where he subsequently lived for most of his life. As it happened, this was the same year that Jefferies, pursuing suburban journalism, was staying in Sydenham, the following year moving to Surbiton, where he lived until 1882. Thus, it was a possibility that the two men might have met, although it seems they never did. Hudson, however, describes what he implies as a ghostly encounter while he was staying at 'Sea View' the house at Goring-by-Sea where Jefferies died. Hudson was somewhat qualified in his praise for Jefferies, feeling that he had not reached his fullest maturity as a writer, and that, had he lived 'he would have written a book about the [South] Downs and the maritime district of Sussex as good as any work we have had from him, I feel certain'.[29] This book, Hudson felt, had been left to him to write, and it was here that he felt he encountered the spirit of Jefferies in human form, while walking in the churchyard near his home: 'I suddenly looked up, and behold, there before me stood the man himself, back on earth in the guise of a tramp! It was a most extraordinary coincidence that at such a moment I should have come face to face with this poor outcast and wanderer who had the Jefferies countenance as I knew it from portraits and descriptions.'[30] He describes the 'long thoughtful suffering face, the long straight nose, flowing brown beard, and rather large full blue eyes'. He gave the man a penny in answer to his request, but did not look him in the face again, 'for I knew that those miserable eyes would continue to haunt me'. Hudson committed the account to paper almost immediately, writing it down 'in the room that was his ... the morning sun filling it with brightest light, the sounds he listened to coming in at the open window – the intermittent whispering of the foliage and the deeper continuous whisper of the near sea, and cries and calls of so many birds that come and go in the garden'.[31] There is a strong sense in this writing that the power of suggestion within the place that conjured the transfiguration of the tramp into Jefferies himself within the mind of Hudson, also communicated a spirit between the two men, held there in this description.

[29] R. Jefferies, *Nature in Downland* (London: Dent, 1923), p. 14.
[30] Ibid., p. 15.
[31] Ibid.

In 1915, Hudson wrote of a windy spring day in Savernake Forest, near Marlborough, that had something of the spirit of Jefferies about it. In it, as his listening grew more intense, more active and concentrated, the feeling of what he heard became almost mystical:

> The sea-like sounds and rhythmic volleyings, when the gale is at its loudest, die away, and in the succeeding lull there are only low, mysterious agitated whisperings; but they are multitudinous; the suggestion is ever of a vast concourse – crowds and congregations, tumultuous or orderly, but all swayed by one absorbing impulse, solemn or passionate.[32]

Another listener, in that very place, might have interpreted the sounds differently, or indeed might not have noticed them in such detail at all. We are like radio receivers, hearing the same signals but playing them out mentally in higher or lesser quality. As his senses tune deeper into the wood, Hudson notices perspective:

> Through the near whisperings a deeper, louder sound comes from a distance. It rumbles like thunder, falling and rising as it rolls onwards; it is antiphonal, but changes as it travels nearer. There is no longer demand and response; the smitten trees are all bent one way, and their innumerable voices are as one voice, expressing we know not what, but always something not wholly strange to us – lament, entreaty, denunciation. Listening, thinking of nothing, simply living in the sound of the wind, that strange feeling which is unrelated to anything that concerns us, of the life and intelligence inherent in nature, grows upon the mind.[33]

Here is Hudson as a sound recorder, with the added quality of analysis as he listens, as did Thomas Hardy in our first chapter, preserving in his poetic prose not only the sound of the forest as the wind buffets it, but also the feelings engendered by being present in the moment, in the place. Reading him, we *hear* the trees moving and empathize with the emotion of his sense of being there.

In the last pages of his strongly autobiographical 1921 book, *A Traveller in Little Things*, Hudson was prompted to turn to actual poetry, in verses inspired by an event in the New Forest, Hampshire. It was here, perhaps, that he came closest to the state to which Richard Jefferies aspired, and of which Jefferies wrote so luminously. Hudson recounts the experience: 'When walking there

---

[32] W. H. Hudson, 'Spring in Savernake Forest' In *Birds and Man* (London: Duckworth, 1927), p. 76.
[33] Ibid.

one day, the loveliness of that green leafy world, its silence and its melody and the divine sunlight, so wrought on me that for a few moments it produced a mystical state.'[34] It was a moment of transfiguration, producing in him 'the idea that we are in communion with or in the presence of unearthly entities'.[35] The poem, entitled 'The Visionary', is in two parts; the first suggests that forces or beings pass through space and time, manifesting themselves when landscape and human blend in a heightened state:

> When on the woodland falls
> A sudden hush, and no bird sings;
> When leaves, scarce fluttered by the wind,
> Speak low of sacred things[36]

The second part of the poem describes his experience as he walks through the forest, listening to birdsong, and gently humming to himself; the wood becomes transfigured in 'radiance white', the grass like 'tongues of emerald flame'. Enraptured by the vision, he sinks to his knees. The moment passes, the Forest resumes itself:

> And pale at first the sunlight seemed
> When it was gone; the leaves were stirred
> To whispered sound, and loud rang out
> The carol of a bird.[37]

We have Darwin and Wallace, the scientists who travelled an epic global journey in search of meaning, and Hudson the knowledgeable ornithologist with a poetic sensibility, adopting as his home the UK and thereafter treating all species – including human – as subjects for detailed observation and comment. Wallace, the sharer in the discovery of the theory of evolution, later developed an interest in Spiritualism, exhorting his followers to apply the same analytical stringencies to the study of this as he had with his natural science work: 'We write not to convince, but to excite enquiry. We ask our readers not for belief, but for doubt of their own infallibility on this question.'[38] Meanwhile there is Hopkins, the Jesuit poet, who needed no convincing, and for whom the sound of nature, the *inscape* and *instress* of everything, was a manifestation of God. Then we have Jefferies, a

---

[34] W. H. Hudson, *A Traveller in Little Things* (London: Dent, 1921), p. 256.
[35] Ibid.
[36] Ibid., p. 257.
[37] Ibid.
[38] A. R. Wallace, *My Life: A Record of Events and Opinion* (London: Pantianos Classics, 1908), p. 196.

man of deep sensitivity and spirituality who saw, heard and felt a presence in the world which he attempted to capture in poetic prose, whilst also working for a more populist market. In his 'Autobiography of a Soul' as he called it, *The Story of My Heart*, Jefferies took himself on an ecstatic personal spiritual journey, of which he wrote, in a third-person note: 'He considers the idea of deity inferior, and believes that there is something higher.'[39] The resulting book divided opinion, as it still does. Hudson felt that, perhaps had Jefferies lived, 'that strain of intense unnatural feeling, which he so strangely misinterpreted, and which in the book just named touches the borders of insanity, would have been outgrown'.[40] On the other hand, Edward Thomas, in his 1909 biography of Jefferies considered that 'as an autobiography it is unsurpassed because it is alone. It is a bold, intimate revelation of a singular modern mind in a style of such vitality that the thoughts are as acts, and have a strong motive and suggestive power.'[41]

What binds all these writers together is a sense of wonder at a strange beauty, touched on time and again in the sounds transmitted from the sometimes mysterious natural world. Darwin himself listens intently and hears sounds as quiet as the movement of sand,[42] or a spider's 'singing', as mentioned in *The Descent of Man*, where he refers to the male Theridion spider, several species of which 'have the power of making a stridulating sound' that have attracted the attention of a number of arachnologists and led to the belief that 'spiders are attracted by music'.[43] In the minutiae lies the meaning – and of course further questions and mysteries – but all with a beauty at the point where science, landscape and poetry meet. Such detailed observation contains the passion of attention; here is Hudson, writing with his own form of passion in the Somerset hills:

> The continuous singing of a skylark at a vast height above the green, billowy sun and shadow-swept earth is an etherealised sound which fills the blue space, fills it and falls, and is part of that visible nature above us, as if the blue sky, the floating clouds, the wind and sunshine, has something for the hearing as well as for the sight. And as the lark in its soaring song is of the sky, so the wood wren is of the wood.[44]

---

[39] R. Jefferies, *The Story of My Heart* (London: Longman, Green, 1936), p. xiv.
[40] W. H. Hudson, *Nature in Downland*, p. 14.
[41] E. Thomas, *Richard Jefferies: His Life and Work* (Boston: Little, Brown, 1909), p. 323.
[42] I cite this instance in a previous book, *Sound at the Edge of Perception* (Singapore: Palgrave MacMillan, 2019), p. 82.
[43] C. Darwin, *The Descent of Man*, p. 315.
[44] W. H. Hudson, *Birds and Man* (London: Duckworth, 1927), p. 103.

Meanwhile Richard Jefferies, ill and close to death at the age of just thirty-nine, and confined to his home at Goring-by-Sea, seemed to be coming to a bitter reckoning as he gazed and listened through his window:

> The old, old error: I love the earth, therefore the earth loves me ... I am Man, the favoured of all creatures. I am the centre, and for me all was made ... I thought myself so much to the earliest leaf and the first meadow orchis – so important that I should note the first zee-zee of the titlark – that I should pronounce it summer ... They manage without me very well ... They go on without me.[45]

Jefferies died on 14 August 1887; W. H. Hudson died on 18 August 1922. The two men are buried a few yards apart, in Broadwater Cemetery, Worthing, in Sussex. On 11 May, a few months before Hudson's passing, radio station 2LO made its first transmission, prior to the start of broadcasts from the British Broadcasting Company on 18 October, from which time the air would become peopled by the sounds, not only of nature, but also of human voices and music, witnesses to 'live' experience and the recounting of memory. For many, the coming of this invisible medium brought its own sense of 'das unheimliche' as sounds reached them through the ether. When, on the wireless in 1924 a nightingale's song was heard, with the cello of Beatrice Harrison in her Surrey garden, it was as though two worlds almost touched, just as they seemed to do when Gilbert White and Henry Thoreau had listened to the ghostly moaning and murmurs of owls in nearby woods. In the meantime, the world of nature, and the idea of its sound, remained ever-transmittable through words, and the continuation of this tradition through writers like Edward Thomas into the twentieth century takes us to new voices heard, recorded and interpreted in words both spoken and written. Thomas himself in his biography of Richard Jefferies, (dedicated to W. H. Hudson,) offers us an entr'acte in celebration of literature and its power as a recording device; speaking of Jefferies's last work, he writes:

> It is the supreme proof, above beauty, physical strength, intelligence, that a man or woman lives. Lighter than gossamer, words can entangle and hold fast all that is loveliest and strongest, and fleetest, and most enduring, in heaven and earth ... The words remain, and though they also pass away under the smiling of the stars, they mark our utmost achievement in time. They outlive the life of which they seem the lightest emanation – the proud, the vigorous, the melodious words

---

[45] R. Jefferies, 'Hours of Spring' in *Field and Hedgerow*, S. J. Looker (ed.) (London: Lutterworth Press, 1948), pp. 22–3.

> ... the things are forgotten, and it is an aspect of them, a recreation of them, a finer development of them, which endures in the written words.[46]

With those thoughts in mind, we move to our final chapter, and the continuing vital necessity of written words as witnesses to sound within our own traumatic and changing time. There would be growing concern and anger at the plight of the planet, and humankind's effect upon its ecology. At the same time, the underlying common ground throughout would continue to be that of a celebration of the sounds of the natural world, expressed both in written text alongside the direct transmitter of an increasingly sophisticated medium of communicating sound recordings. And through it all there remains Darwin's final words in his great book, of joy and affirmation:

> There is grandeur in this view of life, with its several powers, having been originally breathed into a few forms or into one; and that, whilst this planet has gone cycling on according to the fixed law of gravity, from so simple a beginning endless forms most beautiful and most wonderful have been, and are being, evolved.[47]

---

[46] E. Thomas, *Richard Jefferies: His Life and Work*, p. 327.
[47] C. Darwin, *The Origin of Species* (London: Penguin Classics, 1985), p. 460.

11

Listening to ourselves listening: Voices for today and tomorrow

Preserving the exact essence of an original experience, be it an event, an image or a sound, in another form is complex and problematic, because that essence may mean different things to different people, shaped by personal imagination. 'Field recording, composition and poetic text are three forms of writing', comments Michael Pisaro: 'Each form of writing already involves a kind of translation'.[1] A sound recordist, he argues, translates vibration into electrical impulses and thence back into vibration, the whole process translating the earth's original text into a held experience. Likewise, a composer might notate their interpretation of the same sonic 'scene', which in turn is translated back into sound by musicians, to be experienced by the listening audience. Meanwhile, writes Pisaro, 'the written text also represents sounds that have been translated by the mind of the reader into actual or virtual sound. An image described in a text is sketched with a degree of abstraction that asks the reader to fill in the spaces'.[2] Yet in the end, what we are experiencing when we listen to a bird, or a seascape, or trees in a wood is untranslatable, a text beyond language, and we appropriate it through a transference of that experience from the event to ourselves; because we are humans, we seek to humanize what is, when all is said and done, so near and yet so far from who we are.

In March 1857, a Frenchman, Édward-Léon Scott de Martinville (1817–79) patented a device that could transcribe sound waves onto a blackened sheet of paper, showing as an undulating line. The idea was that sound could be made visible, aiding its study. He called his invention the *Phonautograph*, and his recordings, 'phonautograms'. With this machine, he demonstrated that sound could be 'frozen' as a visual artefact. Richard Menke has written of the

---

[1] M. Pisaro, 'Rubies Reddened by Rubies Reddening' In *Writing the Field: Sound, Word, Environment*, S, Benson and W. Montgomery (ed.) (Edinburgh: Edinburgh University Press, 2018), pp. 187–8.
[2] Ibid.

concept: 'A machine for listening, his device was supposed to mimic the human ear ... and then to act as an automized hand, turning sound into sight ... It converted the passage of time from the guarantor of a sound's evanescence into the horizontal timeline that gave it continuity and visual unity'.[3] It was a printed text, but it was also *sound*. At the time, de Martinville had no notion that his 'notation' would ever be playable: it was not his intention after all, and anyway, the technology for playing back recorded sound was still thirty years away when he created his machine. Nevertheless, in 2008, by optically scanning some of the files, and using a computer to translate these scans into digital audio files, the First Sounds project DID succeed in bringing some of these sounds back from the paper; printed vibrations were activated and in one now famous result, the voice of de Martinville himself, singing a few bars of 'Au clair de la lune' could be made out.[4] Other files followed, and 'suddenly, as listeners, we were eavesdropping on a past whose sounds it had never been possible to hear, listening to vibrations set down at a time when Edison, Alexander Graham Bell, and Emile Berliner were children'.[5] A form of recording technology, created when John Clare was still alive, had taken 150 years to gain its voice. Had the capacity for playback been available in the late 1850s, who knows what possibilities might have presented themselves? As it was, in the meantime, words as expressions of human response to sound transferred onto the pages of a book, continued to hold the music of the world for posterity, as they had for centuries.

In 1909, when his biography of Richard Jefferies was published, Edward Thomas was living at Berryfield Cottage, Ashford, near Petersfield in Hampshire. That same year, he published another book, *The South Country*, a chronicle of twelve months' wanderings across England south of the Thames and Severn and east of Exmoor. In its pages, it is not hard to sense the influence of Jefferies, as in this passage, from the chapter on spring:

> On every hand there is a drip and gush and ooze of water, a crackle and rustle and moan of plants and trees unfolding and unbending and greeting air and light ... And over and through it a cuckoo is crying and crying, first overhead,

---

[3] R. Menke, *Literature, Print Culture, and Media Technologies, 1880–1900* (Cambridge: Cambridge University Press, 2019), p. 4.
[4] First Sounds was formed in 2007 by David Giovannoni, Patrick Feaster, Meagan Hennessey and Richard Martin to facilitate and coordinate the recovery of the oldest known sound recordings.
[5] R. Menke, *Literature, Print Culture*, p. 4.

then afar, and gradually near and retreating again. He is soon gone, but the ears are long afterwards able to extract the spirit of the song, the exact interval of it, from among all the lasting sounds, until we hear it as clearly as before, out of the blue sky, out of the white cloud, out of the shining grey water. It is a word of power – cuckoo![6]

There is a sense of presence, of the surround-sound of nature, that is so like Jefferies at his intense best. Much has been said – and rightly so – of the later influence of Robert Frost on the poetry of Edward Thomas, in particular in relation to the sound of sense in language. It is clear, however, that from earlier in his life, his ear was tuned to an extraordinary degree, working with his imagination to seek out layers below layers of signals, where light and sound blend together, working on the imagination to create an overall feeling of transcendence. It is hardly surprising then, that from the very beginning, the writings of Richard Jefferies should find a kindred ear in his reading eye. As Jem Poster has said, 'when Edward Thomas wrote his biography of Jefferies, he was paying homage to a writer who had influenced him, at the deepest level, from childhood onwards … It's clear that Jefferies's works played a crucial role in Thomas's intellectual and imaginative development, and that Thomas's admiration for his predecessor constituted a form of filial devotion.'[7] For Thomas, listening was a key factor in unlocking the world. How could it be otherwise, given the influence of Jefferies, a man for whom micro-listening was a part of natural observation, an audio window on sensation and meaning. From 1913, when Edward Thomas's poems started to emerge, he was able to take his listening into new realms, where memory and the senses combined to open dark corridors not unlike those spoken of by Keats in an earlier chapter. Take his 1915 poem, 'The Unknown Bird', which begins:

> Three lovely notes he whistled, too soft to be heard
> If others sang; but others never sang
> In the great beech-wood all that May and June.
> No one saw him: I alone could hear him
> Though many listened. Was it but four years
> Ago? or five? He never came again.[8]

---

[6] E. Thomas, *The South Country* (London: Hutchinson, 1984), p. 41.
[7] J. Poster, 'Excavating the Future, Richard Jefferies and Edward Thomas: A Spiritual Affinity in Poetry and Prose' in *Times Literary Supplement* (London, 15 June, 2018), pp. 18–19.
[8] E. Thomas (ed.) and E. Longley, *Edward Thomas: The Annotated Collected Poems* (Tarset: Bloodaxe Books, 2008), p. 55.

The poem takes us with Thomas on a journey to a place we cannot quite reach. It does for listening, what his great poem, 'Old Man' does for the olfactory sense. The familiar is tinged with the uncanny, and there is strangeness and mystery in the beech wood. Birdsong at times for Thomas becomes almost a language, but a language he approaches as we might hear foreign voices. It is sound that moves beyond meaning to something purer, communication through intonation, pitch and pause, a form of music that contains an aesthetic rather than a factual truth and which we must accept on its own terms. It may be this quality in his writing that led the well-known ornithologist, writer and natural history broadcaster, James Fisher, to reach the conclusion that he was 'the major English poet of our [i.e., the 20th] century'.[9] Certainly, the final stanza of his poem, 'Sedge-Warblers', written in May 1915, combines the sense of a door opening, leading to somewhere unknowable beyond and at the same time, within us, while adding a dig in the final line against formal learning as opposed to the mysterious wisdom of the natural world:

> Sedge-warblers, clinging so light
> To willow twigs, sang longer than the lark,
> Quick, shrill, or grating, a song to match the heat
> Of the strong sun, nor less the water's cool,
> Gushing through narrows, swirling in the pool.
> Their song that lacks all words, all melody,
> All sweetness almost, was dearer then to me
> Than sweetest voice that sings in tune sweet words.
> This was the best of May – the small brown birds
> Wisely reiterating endlessly
> What no man learnt yet, in or out of school.[10]

Just as the unknown bird's song in the previous poem was beyond received understanding, the humble sound of the sedge-warblers challenges us with something deeper than worldly knowledge. Edward Thomas's poetry and prose are brushed by the sense of the transient, the fleeting sound, smell or feeling. We need to continually remind ourselves that today we are gifted with the ability to be able to play back auditory experience, a facility generally available to very few within the general public during the time Edward Thomas was writing, (although home-recording cylinder kits began to become available during the

---

[9] J. Fisher, *The Shell Bird Book* (London: Ebury Press and Michael Joseph, 1966), p. 208.
[10] E. Thomas, *Annotated Collected*, p. 91.

First World War). Notwithstanding the unexplored potential of de Martinville's *Phonautograph*, to 'listen again' in Thomas's day meant to turn back the page of the book, or to remember. Wars, however, always advance technology, and the one that killed Edward Thomas in 1917 indirectly led to the creation of mass communication that would ultimately bring the sounds that were beginning to be recorded by specialists such as Ludwig Koch, into the living rooms of the world. Fully portable recording, however, was still distant. It would take another war before real freedom of movement for sound recordists would be developed, as journalists followed allied armies across Europe in pursuit of Hitler's retreating forces. Today, we possess the tools to remind ourselves of past sounds, and to examine them at our leisure. It is salutary to understand that some of the instruments through which we celebrate the sounds of the world, have themselves been developed at a cost to human life and the environment. David George Haskell pointed this out in his book, *Sounds Wild and Broken*, through the example of simply playing a vinyl record of natural sounds, a diamond stylus tracking through the grooves of a polyvinyl chloride disc: 'The jewel follows the wavy plastic groove, every microscopic side-to-side motion conveyed to magnets and wire coils in the stylus's head. Burned coal and methane, arriving on wires strung across the sky, electrify my amplifier'.[11] We would not be without our precious recordings, but it is as well to remember the irony of technological development.

So why should we still turn to the written word as a means of 'recording' the audio experience of the field? (After all, the book itself has been found guilty of contributing to deforestation). The explanation lies in the answer to a question I put to the sound recordist Chris Watson, and it brings us back to the point discussed several times in this book, that of the necessity for human expression to communicate both the event itself and the personal experience of witnessing it. Over a number of years, Watson issued a range of commercial CD recordings, accompanied by texts as comprehensive as the format allowed within the physical framework. Thus, he was effectively revisiting the concepts of the sound-books of Ludwig Koch that opened doors for many of a previous generation, before audio portability became ubiquitous:

> I was forced to be brief because they were sleeve notes so I couldn't write pages … I wanted to get across two things: the technique, because that was part of the

---

[11] D. G. Haskell, *Sounds Wild and Broken* (London: Faber & Faber, 2022), p. 291.

process, and the feelings – what I felt, both during the recordings and listening back to them. The two were inseparable even though they come from two different places, the technical and the emotional.[12]

Watson's words remind us of the necessity of reflecting on the experience, and recording how it makes us feel. Text continues to be complementary to the heard sound, offering the opportunity of description of both the sounds we hear and the experience we preserve. Hence, Tim Dee can write of a Skylark on Tubney Fen letting out 'a shivery call as it rose', and raising a song that began, 'spooling out from those first few gargled notes', and then a few lines further on noting that 'a crane called: one brass blast, stretched and bent, as if from an old fog horn on the fen … It could have been the air itself being hurt, the sky cracking'.[13] Dee has worked as a radio producer, a naturalist history writer and a poet, and here the disciplines become one in the precision and originality of sound finding expression in words. Locating the phrase to describe the sound so exactly is much more than a pleasing aesthetic piece of literary expression. It is part of listening, and it has a practical purpose too; for those of us who may not be trained ornithologists, finding within ourselves a means to describe birdsong, or indeed any sound phenomena, helps to root the identity of the experience and its source in our memory, so that it becomes a part of learning, understanding and belonging to the world ourselves. Here is Edward Grey in his 1927 book, *The Charm of Birds*, listening to a pair of goldcrests, 'the little call-notes … like needle-points of sound'.[14] To listen more deeply, and allow the mind and hearing to become, as it were, at one with the sound is, as Pauline Oliveros writes, to enter a new layer of attention:

> There are two modes of attention: 1) focal attention, which corresponds to the all-or-nothing state – attention to one point and nothing else, and 2) global attention, which corresponds to an open receptive state – attention expanded to a field. Focal attention is sharp and clear. Global attention is warm and fuzzy. The two modes work together as expansion and contraction.[15]

---

[12] C. Watson, quoted in S. Street. *The Poetry of Radio* (Abingdon: Routledge, 2014), p. 104.
[13] T. Dee, *Greenery: Journeys in Springtime* (London: Jonathan Cape, 2020), p. 60.
[14] E. Grey. Viscount Grey of Falloden. In *The Charm of Birds* (London: Hodder and Stoughton, 1929), p. 38.
[15] P. Oliveros, *Sounding the Margins: Collected Writings 1992–2009* (Kingston, NY: Deep Listening Publications, 2010), p. 29.

By moving from global to focal attention, Grey, listening to the gold-crests, finds the words that work for him as a mnemonic to call back the sound in another location and time: 'It suggests to me,' he writes, 'a tiny stream trickling and rippling over a small pebbly channel, and at the end going over a miniature cascade'.[16] There is his recording, his personal sonic text that will aid him to remember, and to identify the bird the next time he hears it in amongst the other voices of a landscape or a garden. We have to listen in our own way if we are to own the sound as part of our living experience. Grey seems to have taught himself to listen almost as the bird itself listens; hearing two wrens engaged in what he calls 'song combat', he notes the listening pauses, the call and response of the dialogue. He is by a greenhouse in a busy garden, there are people about, but the wren is oblivious to all but the unseen rival. 'The strophe and antistrophe went on: the attitude of my wren when listening was intent and still; when it replied the animation and vehemence were such that it seemed as if this little atom of life might be shattered by its own energy'.[17]

As Oliveros indicated, the mind and ear, when working in synchronicity, can switch from micro to macro in a moment. Nan Shepherd's book, *The Living Mountain*, describing her journeys through the Cairngorm Mountains, devotes a whole chapter to the senses, noting 'the noises birds make that are not singing, and the small sounds of their movements' at the same time as experiencing the chaos of a surrounding storm:

> Gales crash into the Garbh Choire with the boom of angry seas: one can hear the air shattering itself upon rock. Cloud-bursts batter the earth and roar down ravines, and thunder reverberates with a prolonged and menacing roll in the narrow trough of Loch Avon. Mankind is sated with noise; but up here, this naked, this elemental savagery, this infinitesimal cross-section of sound from the energies that have been at work for aeons in the universe, exhilarates rather than destroys.[18]

We are, in general terms, a more consciously visual society than an auditory one, having lost some of the survival instincts that made our ancestors tune to the world around them. This is not to say that we cannot train ourselves in the art of listening, as in Oliveros's mantra: 'Listen to everything all the time and remind yourself when you are not listening'.[19] Meanwhile, the world at large changes,

---

[16] E. Grey, *The Charm of Birds*, p. 39.
[17] Ibid., pp. 10–11.
[18] N. Shepherd, *The Living Mountain* (Edinburgh: Canongate Press, 2008), p. 97.
[19] P. Oliveros, *Sounding the Margins*, p. 28.

and for the most part, we run the risk of becoming desensitized as much to the warnings, as the joys of the environment. On the other hand, when true survival is at stake, there is something in our power of attention that switches on in spite of ourselves. The fact is that, given certain circumstances, we still have the capacity within us to listen like an animal or a bird. In 2018, a book called *The Salt Path* by Raynor Winn became a best seller. It is an account of Winn, and her husband Moth, who had been diagnosed with a terminal illness, as they set out on a 630-mile odyssey along the South West Coast Path of England. On the journey, having lost virtually all their worldly possessions, including their home through bad investments, they tune themselves back to fundamentals, notably in response to nature, wild camping and simply living life by setting one foot in front of another. The book opens with a moment of danger; they have pitched their tent on a beach above – as they believe – the tide line. Raynor is on the edge of sleep, but something in her auditory subconscious alerts the drowsing senses:

> There's a sound to breaking waves when they're close, a sound like nothing else. The background roar is unmistakeable, overlaid by the swash of the landing wave and then the sucking noise of the backwash as it retreats. It was dark, barely a speck of light, but even without seeing it I recognized the strength of the swash and knew it must be close. I tried to be logical. We'd camped well above the high-tide line; the beach shelved away below us and beyond that was the water-level.: it couldn't reach us; we were fine. I put my head back on the roll-up jumper and thought about sleep. No, we weren't fine, we were far from fine. The swash and suck wasn't coming from below, it was right outside.[20]

We listen, not only in stereo, but also with perspective. Near and far sounds: we may not articulate to ourselves how we distinguish them, but we do, and they prepare us for what comes next, what is approaching, or receding, or passing us. Interpreting them, we build anticipation, expectation, or fear and the need for action. Stepping outside with a critical ear that interrogates sound, we begin to understand what it is to listen beyond our species. The ability to articulate this, however, is another matter.

There are books which we consider as almost contemporary that are quite rightly, already classics, such as J. A. Baker's *The Peregrine*, that belong to this realm. Baker was an office worker, employed by the A. A. in Chelmsford, Essex, who happened to be fascinated by raptors, and his book, based on months of

---

[20] R. Winn, *The Salt Path* (London: Penguin, 2018), p. 1.

observations, has been acclaimed as a masterpiece of the twentieth-century nonfiction. His work is at once a corrective to any over-sentimentalized view of the natural world in literature that might creep in as we celebrate it, and at the same time an individual voice that speaks from a personal compulsion and passion. His was a human response to a bird that seared itself into his consciousness as if he had come to share the same sense of meaning; there are times in the book where the words seem to emerge out of the bleakness of wild places like the creatures that inhabit them.

Whether we are weekend visitors, part-time observers or country born and bred residents, whether we are the blind Edward Rushton hearing an eighteenth-century robin at Liverpool Docks, or John Alec Baker cycling into the Essex countryside on his day off from his Chelmsford office, in search of the killing fields of the peregrine, we are listening-in from another room. In the best hands and minds, this can be appropriated in the service of understanding. Baker's obsessive attention to avian life – and particularly death – in the Essex landscape mostly to the east of Chelmsford, is almost literally twittering with sound, and his writing is of an animalistic intensity. It is one thing to listen, but it is another to communicate that listening, to make a subconscious act conscious, and share it as if first hand. We learn that, in Baker's words, 'for a bird, there are only two sorts of bird: their own sort, and those that are dangerous'.[21] When the falcon is hunting, even when it is too high to be seen by the naked human eye, the air itself becomes electric with danger, and we hear the sounds of the landscape change as it becomes alarmed. Visceral as the killing scenes are in *The Peregrine*, there is also great beauty; for example, early in March, 'all day the unquenchable larks sang. Bullfinches lisped and piped through the orchards'. And the next day, as morning breaks: 'owls were calling in the long dim twilight before dawn. At six o'clock the first lark sang, and soon there were hundreds of larks singing up into the brightening air. Straight up from their nests they rose, as the last stars rose up into the paling sky'.[22] Then, out of this perfect morning, death comes, and Baker teaches us that we cannot have the beauty without the violence. As the day goes on, he sees a merlin on the hunt:

> It flew forward into the wind towards a skylark singing high above the fields. It had seen the lark go up, and had circled to gain height before making an

---

[21] Baker, *The Peregrine* (London: William Collins, 2017), p. 56.
[22] Ibid., p. 149.

attack ... It reached the lark in a few seconds, and they fell away towards the west, jerking and twisting together, the lark still singing ... Their rapid, shifting, dancing motion had been so deft and graceful that it was difficult to believe that hunger was the cause of it and death the end.[23]

Writers such as Baker bridge science, observation and a poetic sensibility in their description, but no matter how skilled, a written text will not hold the *actual* sound. It can, however, convey an emotional relationship. Early in *The Peregrine*, there are two sentences that encapsulate the nature of capturing the sound of the natural world in words: 'I do not believe that honest observation is enough. The emotions and behaviour of the watcher are also facts, and they must be truthfully recorded'.[24] It may be a symptom of the times in which we live that the years at the end of the twentieth century and the first decades of the twenty-first have produced some of the most distinguished writing about the natural world ever in both poetry and prose. As for John Alec Baker, he aspired to be a poet, and his book is as much prose poem as it is natural history. It is also full of anger against human destruction of the natural world; at its heart is not just a bird or a landscape, but the relationship between a human being and both.

Sound recordists who are experienced with wild audio likewise bring a necessary knowledge and understanding to their task; it is more than a question of listening: you need to know, or at the very least, try to understand, what you are listening to and what its implications are. Bernie Krause is one of the world's leading experts on natural sound, and has written eloquently on the process of capturing it. His account of a dawn chorus in Zimbabwe is memorable for its ability to demonstrate his approach to recording, and his response to the location and its acoustic qualities, the variety of sound within it. In order to do so, he turns to a musical analogy: 'The acoustic moment was so rich with counterpoint and fugal elements that it immediately brought to mind some of the intricate compositional techniques used by Johann Sebastian Bach (as in his Prelude and Fugue in A Minor)'.[25] We are used to the attempts of composers as they seek to mimic birdsong, but here is something different, an unconscious sense of kinship rooted in common ground – or rather air. What Krause also notices, is that the climate and environment had produced a kind of sound space similar to a man-made recording studio: 'The birds and insects were vocalising in a habitat

---

[23] Ibid., pp. 150–1.
[24] Ibid., p. 32.
[25] B. Krause, *The Great Animal Orchestra* (London: Profile Books, 2012), p. 28.

that had seen no rainfall for weeks, and there were no reflective surfaces in this environment and thus no echoes'.[26] At the same time, there were baboons high above on a granite outcrop that sent their voices reverberating down and into the woodland, 'the acoustic delay lasting for six or seven seconds before fading into silence. It was a unique gathering of sounds specific to this singular place, and their voices created an eerie imbalance within the soundscape – the dry, non-reverberant sound of many birds and insects set off against the long-echoed voices of the few baboons'.[27] At the centre of Krause's writing is the desire and ability to listen and to attend: 'If we wish to hear creature orchestras again, we have to slow down, become very quiet, and listen intently'.[28]

Recording in the field involves analysing the dynamics of the sound world as it emerges, passes and fades, and it produces its own vein of philosophy. The recordist, broadcaster and sound artist Geoff Sample has written illuminatingly of the shape and structure of a dawn chorus, a layered music that builds upon itself as individuals within a local community of birds enter the group conversation 'until there is a period from maybe half an hour to an hour after the first robin, leading up to the sunrise, when the majority of all species are singing'.[29] Sample speaks of 'species layers', waves of sound that come and go after sunrise, with some individuals taking the lead as soloists from time to time: 'Within these species layers is a spatio-temporal spread of similar sounding voices and similarly structured utterances (the species song); but since neighbouring individuals of any species usually share at least some of their repertoire of sounds or motifs, there is a sense of patterns echoing into the distance, cascading like a fractal into finer detail'.[30]

It is not by chance that many of the best sound recordists close their eyes in order to gain focus. The visuals within the natural world are seductive, and in order to fully comprehend, we must move from what Oliveros has called 'primary or initiatory listening',[31] that is to say from the innocent hearing that we knew when we entered the world for the first time, to a category of attentiveness which we might call interrogatory, the beginning of true understanding.

---

[26] Ibid., p. 29.
[27] Ibid.
[28] B. Krause, *Wild Soundscapes: Discovering the Voice of the Natural World* (New Haven: Yale University Press, 2016), p. 23.
[29] G. Sample, 'Choral Twitter' In *Songs of Place and Time: Birdsong in Natural History and the Arts*. M. Collier (ed.) (Manchester: The Gaia Project/Cornerhouse Publications, 2020), p. 216.
[30] Ibid.
[31] P. Oliveros, *Sounding the Margins*, p. 248.

A child's ability to listen can teach us much; close our eyes and attend. Few writers of fiction have demonstrated the transference of sound to the page through the imaginative hearing of the very young, better than the children's author, Lucy M. Boston. In her series of *Green Knowe* novels, which she based on her own home at Hemingford Grey, Cambridgeshire, Boston places herself in the world of a young boy, Tolly, cohabiting an ancient house with spirit children who come and go through sound, occasional glimpses and place-memory. In the process of her narratives, Boston plays the imaginative sound of this environment, aiming it at the ears and minds of her youthful readership. In the second book of the series, *The Chimneys of Green Knowe*, Tolly learns the story of a girl called Susan, who lived in the house during the English Regency period. He discovers that Susan was blind, and he sets himself to sense the world as she would have done, through listening with his eyes closed, and by so doing, he telescopes the time between them. Immediately, sound becomes heightened as he lies on a grassy bank, trying to listen to the wind, experiencing it as Susan might have done:

> At once [he] realised how much wind there was, and how big, tell-tale and friendly. It bumped into and passed round sheds, it crossed the gravel, bowling protesting leaves before it. It made a different sound in each tree, in some like the sea, in others like fretted tissue paper. How it whirled the yew branches about! Tolly could imagine the clouds moving like ships under full sail, but Susan would know nothing about clouds or sky, and never could. But surely somehow she would feel the *size* of the wind? She could hear its approach from far away and the immense hubbub of its passing. The birds were trying to sing, because it was March, but the song was interrupted, jerked out of their throats as they were tossed off the branches with an extra flutter that reminded Tolly of rowing in rough water.[32]

Boston offers us new ways of seeing and hearing through new ways of saying, finding within the text, sounds to surprise the ear into a personal truth. That is the underlying reality of the human instinct to record, in whatever form, be it through word on paper, or direct sound onto cylinder, disc, tape or memory card; to hold something for memory and the future, to preserve it against time. We do this because we want to understand and/or share that understanding. If we engage, sometimes obsessively, in trying to mimic what we hear around us,

---

[32] L. M. Boston, *The Chimneys of Green Knowe* (Hemingford Grey: Oldknow Books, 2003), pp. 43–4.

it is hardly to be wondered at, because we are participants as well as being at the same time, outside witnesses.

That sense of witness becomes more important as we see the world withering before us, but it has always been there, this need to hold on to the sound of 'how it was when we were there'. The art of true listening is to comprehend the layers of sound beneath sound, and beyond that, the silence. Henry Beston, in a lonely shack amongst sand dunes on Cape Cod during the 1920s, wrote of his landscape/seascape after the birds had gone, left with only 'the long wintry roaring of the sea. Listen to it for a while, and it will seem but one remote and formidable sound; listen still longer and you will discern in it a symphony of breaker thunderings'.[33]

As to birdsong itself, our nightingale mimic from Chapter 1, Madame Saberon, has had a number of successors in the media over the years, many of them radio and recording stars in their own right. Only a few years after her 1924 impersonation, the BBC employed a young plough maker from Suffolk as a bird imitator at its first London headquarters, Savoy Hill. His name was Percy Edwards, and he went on to belong to two worlds, working as a popular entertainer in theatres and on media alongside such as Max Miller, Charlie Chester, Peter Sellers and Morecambe and Wise on one hand, and with experts from the field of natural history broadcasting on the other, among them Tony Soper, James Fisher and David Attenborough. Edwards was ubiquitous in his time, used by radio producers in radio dramas as well as on variety shows, and capable of remarkable mimicry that had grown from close listening and observation. His fame led to a number of books, including volumes of autobiography. He was at root a naturalist; his knowledge was extensive, and his ability to communicate the spirit of a place and the sounds within it demonstrated itself on the page as on the air. In an early book, *Call Me at Dawn*, he painted word pictures of riverside and open field alike in which the ambience of an early-morning postwar (1948) countryside communicated to the reader:

> Here too the curlew may be heard, and of these birds I would say that no song (excepting that of the redshank) is more in harmony with solitude ... Somewhere in the distance behind me a church clock booms six hours, and as the last reverberating note dies away there is a warning *Tu-tu-eu* from the redshank.[34]

---

[33] H. Beston, *The Outermost House* (London: Pushkin Press, 2019), p. 43.
[34] P. Edwards, *Call Me at Dawn* (Ipswich: East Anglian Times, 1948), p. 18.

As the day breaks, more sounds come, and Edwards weaves them into the narrative of his experience in the place, until: 'Now there is a sound like the throbbing of distant machinery. *Wham-wham-wham.* But this is a sound of nature – of the powerful wing-beats of two herons as they fly towards me.'[35]

Through all his work, Edwards shows an understanding that the sound needs a context. In his popular radio broadcasts, he would weave impersonations into a spoken contextual narrative, and his books worked in the same way, although silently, and this is where listening through the ear and the imagination, separate. The experience and the sonic phonetics may be conveyed on a page, but not the pitch, the tone and the inflections of the music itself. As John Bevis observed, 'naturalists have improvised and improved on their own and each other's results, occasionally agreeing, sometimes quite at odds, but more often rendering the same sound in variations that are nitpickingly slight.'[36] Bevis himself created a lexicon of written sounds as an aid to identification, including a section on mnemonics from the realms of folklore. In the end, we must concede that all this is aspiration; birds do it better than we ever shall, because they are made differently. Perhaps we should just save our breath, and focus on learning how to listen.

By finding ways of making the experience of sounds new and fresh, we take our listening to a new level. There are simple ways of doing this. As we insulate our homes against weather and intruders, we also insulate them against air and sound. Simply open a window and the sound of the world comes flooding in. Likewise, activate the sound recorder on a portable device: even a smart phone. Turn the recording level to zero, and put a pair of headphones on, preferably the closed sort, ideally noise-cancelling. Now slowly wind the record level up, listening all the time. Gradually the silence fills with something that was always there beyond it. This new awareness can foster expression, so note down the sensation as the curtain is pulled back. What did it feel like to be there then? Sound is dynamic, it comes out of apparently nowhere, and a song bird as it starts to sing is an event in time. We have focused on birdsong in this book, because it is still, at least for now, available as a sound source. But birds for the most part do not make their sounds for us, they are mostly indifferent to our attention. We may mimic them, and write down approximations of their songs, but they remain separate and mysterious, which is one of the reasons we continue to

---

[35] Ibid., p. 20.
[36] J. Bevis, *Aaaaw to Zzzzd: The Words of Birds* (Cambridge, MA: MIT Press, 2010), p. 31.

listen, seeking to understand. The sounds of the natural world – particularly the animal kingdom – have been our teachers since we first lifted our heads and listened. Darwin acknowledged that fact, and as Morton and Page have written so persuasively: 'Just below the surface, and often right on the surface, naked as a bird's alarm call, can also be heard the cumulative processes that began when some primitive amphibian grunted in the open air'.[37] If we listen – actively listen – we might reach something else, and if it is only a momentary sense of wonder, that, in itself, may actually be the beginning of a new and fuller understanding. 'Beauty is truth, truth beauty.' So the Grecian Urn told John Keats in his famous Ode, to which Keats adds this final thought: 'that is all/Ye know on earth and all ye need to know'.[38] While there are numerous ways of saying it, the fact remains that if our ability to tune our senses – or what lies beneath them – fully to the world, enables only that, it will have been enough. In an essay on listening in the heart of Wales, Jay Griffiths wrote:

> Birdsong seems to happen on the horizon of the human mind, just beyond the extent of our senses. Immanent but untranslatable – the dash –!– the glimpse, the hint, the ellipsis. All birdsong is always partially eclipsed to us, as if it is always leaning towards the leading note, the seventh keening for the tonic, as a skylark, self-leading, rises higher and higher, to the high-octane octave – yet – always – leggerissimo, as lightly as possible, where light is both weight and sound, both brightness and joy, and the octave is reached only at a point of silence created by the very quintessence of its own music.[39]

We seek, and will continue to seek, ways of explaining what we hear, and in so doing attempt to prove worthy of it. Perhaps, however, there is nothing to explain after all. Some would say that it is music, no more, no less, and beyond explanation. David Rothenberg is a philosopher and also a clarinet player; he found that when he played, birds sang with him, just as when Beatrice Harrison played her cello, a nightingale joined her in song, a discovery that led him to write a book, *Why Birds Sing*. As he points out, 'a lovely piece of music actually says nothing at all. Birds certainly sing to find love and to find home, but these reasonable purposes do not deny joy'.[40] If we follow this philosophy, we can relax in the knowledge that the first and most important part of understanding is a

---

[37] E. S. Morton and J. Page, *Animal Talk* (New York: Random House, 1992), p. 262.
[38] J. Keats, *Poetical Works* (London: Oxford University Press, 1967), p. 210.
[39] J. Griffiths, 'Birdsong: Hannah's Wood, Heart of Wales' in Collier, p. 54.
[40] D. Rothenberg, *Why Birds Sing*. London: Penguin Books, 2006, p. xi.

sense of wonder, and so leave ourselves free to listen as partners rather than outsiders, and celebrate.

Yet this celebration must be to some degree muted; in the 2020 spring of Covid, the world locked down and listeners rejoiced in the consolation and benediction of birdsong. What though, if the situation is to be reversed? Rachel Carson had envisaged such a thing in her famous book, *Silent Spring* in 1962, in which she imagined that one day, 'on the morning that once throbbed with the dawn chorus of robins, catbirds, doves, jays, wrens, and scores of other bird voices, there was now no sound; only silence lay over the fields and woods and marsh'.[41] In the end, we can only exhort ourselves towards an aspiration to preserve and protect the world against such a silence. It is as the poet Glyn Maxwell has written: 'What evolutionary psychologists – and I – believe is that aesthetic preferences, those things we find beautiful, originate not in what renders life delightful or even endurable, but in what makes life *possible*'.[42] To record wild sounds onto a memory card should be undertaken with the same level of concentration and care as to focus sounds' meaning in a sonnet. When modern writers seek to express the feeling of sound in the natural world, they are continuing to seek just such precision in the long tradition of authors and poets across time, just a few of whom this book has sampled. The raw material for epiphanies is all around us, happening all the time, often when we least expect it, when our intellectual guard is down. One Christmas Day, on the beautiful Murlough Beach in County Down, the teenage naturalist, Dara McAnulty found himself immersed in such a moment:

> We return to the car park feeling euphoric, breathless. There's a chatter in the trees, but I have to stand very still at the fence bordering the reserve to see into the mist, making out the shapes of twite or linnet, one perched on every bare branch. Little clockwork movements, a chittering chorus. They rise up and land in the field and start busily scuffling at the ground. A song pierces the chatter, as bold as a robin's but this one is a dunnock, throat pulsing with effort, and all this mist around it, lilting with phrasings, bursting through matter. I do one of my wriggly jumps because no one is here to see it.[43]

Dara does 'one of his wriggly jumps', because there is simply no other way to respond to or describe a moment that is so magical, elusive and inexplicable.

---

[41] R. Carson, *Silent Spring* (London: Penguin, 2000), p. 22.
[42] G. Maxwell, *On Poetry* (London: Oberon Books, 2012), p. 10.
[43] D. McAnulty, *Diary of a Young Naturalist* (Toller Fratrum: Little Toller, 2020), p. 189.

The poet and critic Jeremy Hooker (1941–) encountered just such a moment, prompting his poem, 'Somewhere, a blackbird'. Hooker's critical writings have helped us at various points on this journey, but here he articulates the very nub of the issue:

> 1
> Open the door: the song
> sounds at first like a word,
> a lyrical word.
>
> But this is not a word
> we speak or can translate.
>
> Where does it come from?
> Not from the blackthorn
> that is still without a flower.
> From behind the fir, perhaps,
> where magpies at their nest
> taste blood in the song.
> 2
> The words become a stream,
> a musical stream.
>
> But this is air, not water.
> Spirit of air
> with a freshness that blows in.
>
> It opens a space in the mind.
> It discovers a country
> in the winter land.
>
> I cannot explain what it means.
> You have to open the door.[44]

At the heart of it all, there remains a mystery. The ability to listen again, the gift that sound recordists such as Ludwig Koch, Bernie Krause, Geoff Sample, Chris Watson and others have provided, that enables us to pause, play back and recollect sound accurately and in tranquillity, may offer an awakening of conscience. Listening to our recordings of the natural world, the ability to reflect

---

[44] J. Hooker, 'Somewhere, a blackbird' from *Word and Stone* (Bristol: Shearsman Books, 2019), p. 56. Copyright © Jeremy Hooker, 2019. Reprinted by permission of the author and the publisher.

on the witness of what we have heard, the message we have preserved in sound, offers us a chance of true empathy, potentially in the nick of time. But it doesn't explain meaning. We need the words too, because the recorded sounds alone cannot tell what it was like to hear them for the first time, although they can rekindle the memory of those who recorded them. The words will tell the future how it felt in our day, just as they help us to remember. John Berger once wrote that 'a photograph is an automatic record through the mediation of light of a given event: yet it uses the *given* event to *explain* its recording. Photography is the process of rendering observation self-conscious'.[45] We may apply the same definition – with the substitution of 'air' for 'light' – to the process of sound recording, whether it be through a machine or in our mind. Sound recording is the process of rendering *listening* self-conscious; when we take a photograph, we are witnessing ourselves seeing and when we record a sound, we are listening to ourselves listening. Meanwhile, words can help us interpret what we hear, to share with ourselves and others, helping us answer the questions, 'what is it I'm hearing? who is it that listens?' and if ultimately, there is to be a future silence, to fill that silence with memory. The written word – for as long it has existed – has sought to hold on to the ephemeral, confirming and articulating what our senses told us in that moment, a bridge across time, to share with others after the sound – and we ourselves – have gone. It starts and ends with listening.

---

[45] J. Berger, *Understanding a Photograph* (London: Penguin Modern Classics, 2013), p. 19.

Postlude: 'Adieu! adieu! thy plaintive anthem fades'

Some thoughts on the nightingale

We began surrounded by the sounds of a Dorset wood, as evoked in the opening words of Thomas Hardy's novel, *Under the Greenwood Tree*, and we end with a spoiler alert for the ending of that same book. Dick Dewey has married the love of his life, the school teacher, Fancy Day. As they leave the wedding breakfast, and ride away under a moon just over the full, Dick turns to his new bride, sitting quietly beside him:

> We'll have no secrets from each other, darling, will we ever? – no secret at all.
>
> 'None from to-day', said Fancy. 'Hark! what's that?'
>
> From a neighbouring thicket was suddenly heard to issue a loud, musical, and liquid
>
> voice., 'Tippiwit! swe-e-et! ki-ki-ki! Come hither, come hither, come hither!'
>
> 'O, 'tis the nightingale,' murmured she, and thought of a secret she could never tell.[1]

The sound of the nightingale has been a motif throughout our journey, a voice that has made itself heard from the earliest writings of human experience of the natural world. Its beauty has prompted poets, playwrights, naturalists and sound recordists alike to seek it out and set it like a gem in words and soundscapes. It is the most written-about song of all, the voice of a rather plain little bird with an immortal voice. There have been anthologies of poetry, and numerous prose books have been written about it. More often than not, what the words try to catch is the melancholy the mind seems to hear in the music. Why should this be? For the bird itself, it comes down to the old story: males trying to attract females.

---

[1] T. Hardy, *Under the Greenwood Tree* (London: Penguin Classics, 1998), p. 159.

The better the singer, the more support they are likely to offer their young family by feeding and defending them from predators, and the more complex songs, requiring greater effort, show the bird to be in good physical condition. None of which has much to do with the human obsession for claiming the bird's song as a literary icon. Plaintive, yes, melancholy, yes, (although Samuel Taylor Coleridge and D. H. Lawrence disagreed) and beautiful, always, yet there is a darkness in the human story of the bird's song that has to be confronted.

It is a shock for many that at the root of the story of the nightingale's supposed song of sorrow is a reference in one of the most horrific of the Greek myths, coming down to us through the writing of the Roman poet, Ovid. Philomela was the daughter of Pandion, a legendary king of Athens. Her sister Procne married Tereus, king of Thrace, and went to live with him there. After five years, Procne wanted to see her sister. Tereus agreed to go to Athens and bring Philomela back. However, Tereus found Philomela so beautiful that he raped her, then cut out her tongue so she could not tell what had happened. He hid her, telling Procne that her sister was dead. Unable to speak, Philomela wove a tapestry depicting the story and arranged for an old woman to take it to Procne, and when Procne saw the weaving, she asked the woman to lead her to Philomela. After rescuing her sister, Procne planned revenge on her husband. She killed their son Itys and served him to Tereus for supper at the end of which Philomela appeared and threw the boy's head on the table. Realizing what had happened, Tereus chased the women and tried to kill them, but before he could catch them, the gods transformed them all into birds. In the original Greek version, Tereus became a hawk (or a hoopoe), while Procne became a nightingale and Philomela a swallow. Roman writers reversed these roles, making Philomela a nightingale and Procne a swallow. Versions of – and references to – this grisly story have found their way into almost all strands of writing over the past 1,800 years, including the troubadours in France and Spain, and within English literature, Chaucer's *The Legend of Good Women*, Shakespeare's *Titus Andronicus*, John Keats's *The Eve of St Agnes* and even T. S. Eliot's *The Waste Land*.

It is Shakespeare who writes one of the most graphic accounts of the Philomela legend in *The Rape of Lucrece*, when the eponymous central figure, after being raped by Tarquin, compares herself to the Greek maiden: 'Come Philomel; that sing'st of ravishment'. A few lines further on, Lucrece goes on:

And for, poor bird, thou sing'st not in the day,
As shaming any eye should thee behold,

> Some dark deep desert, seated from the way,
> That knows not parching heat nor freezing cold,
> Will we find out; and there we will unfold
>> To creatures stern sad tunes, to catch their kinds.
>> Since men prove beasts, let beasts bear gentle minds.[2]

The darkness of the legend transmits, and speaks to us across time. So perhaps it's all about guilt and conscience? As the late radio features producer, Piers Plowright said, 'music always means more than its sound'.[3]

Hardy also makes a Shakespearean reference to the sound in his coda; with the words 'Come hither … come hither!' he is quoting from a song in *As You Like It*, that also contains the title of his novel:

> Under the greenwood tree
> Who loves to lie with me,
> And turn his merry note
> Unto the sweet bird's throat,
> Come hither, come hither, come hither:
>> Here shall he see
>> No enemy
> But winter and rough weather.[4]

In a note on Hardy's text, Tim Dolin has pointed out that 'if the nightingale's song harkens the lovers back to the pastoral realm, where their only enemies will be winter and rough weather, then Fancy clearly has her doubts, for this bird may be the harbinger of spring, but it does not live long in a cage'.[5] And like Philomela, Fancy Day's tongue cannot tell.

Elsewhere, the song of the bird takes Keats in his famous ode through every emotion, from melancholy, to ecstasy, sorrow at mutability and understanding of birdsong's capacity to speak across time, and in this, he is not alone; as Richard Mabey wrote, 'when Keats called the nightingale the "immortal Bird", he was talking of the way the bird's song seemed to transcend the individual, mortal singer, and how it had been an immemorial antidote to decay and grief. But he could well have been remarking on its extraordinary persistence as a

---

[2] W. Shakespeare, *The Rape of Lucrece*, lines 1128–9/1142–8.
[3] P. Plowright, quoted in S. Street, *Sound at the Edge of Perception* (Singapore: Palgrave MacMillan, 2019), p. 119.
[4] W. Shakespeare, *As You Like It*, Act ll, scene v. Sung by Amiens.
[5] T. Dolin, (n.) in Hardy, *Under the Greenwood Tree* (London: Penguin Classics, 1998), p. 219.

symbol in Western culture.'[6] Sadly, beyond literature, the nightingale seems to be far from immortal; we come to rely on our recordings, written and sonic, and as Bethan Roberts pointed out in her 2021 book on the bird, 'the shrinking of Britain's nightingale population to fewer than five thousand five hundred pairs (a decrease of 93 per cent over the past fifty years) is one of the biggest declines of bird breeding in the UK since records began.'[7] The shock to the system of even reading those words should, to adapt Keats, 'toll us back ... to our sole selves', because, extreme as that statistic may seem, it is also indicative of the urgency attending the need for action to preserve what we have left of the world, and enable future generations to be able to do more than simply read about it or play recordings.

In the meantime, as Sam Lee says, however eloquent the poet, however literally precise the transcription, in the end we remain defeated. The aspiration to hold the sound in a cage of words has produced some of the greatest literary works in the world, but the impetus to write them comes from a frustration that we cannot be as the bird is, and our song can do no more than hint at the reality.

> I realise now that trying to describe the nightingale's song, and the experience around it, is akin to retelling one's previous night's dream to the barista making your morning coffee. It doesn't quite translate. It's a sound that can only truly be understood through first-hand experience, and justice is never quite served when attempting to articulate it ... The truth of the challenge lies in the way the nightingale possesses the strange skill of uttering every possible character tone and textural shape all at once.[8]

We should be content to accept that the song belongs to the nightingale, and so be consoled that we may take the witness of the words with us to places where the bird itself cannot go. If ever there was a poem written with the capacity to transport the mind, the inner eye and the imaginative ear, it is George Meredith's 'Night of Frost in May'. Meredith, we remember, wrote 'The Lark Ascending', the poem which inspired Ralph Vaughan Williams to create one of the most beloved of solo violin-concert pieces. Meredith's hymn to the nightingale is, if anything, even more transcendent in its longing to transport the song of the bird through the silent sound and imagery of the written word – the place and the moment – to wherever we are when we read it or remember it:

---

[6] R. Mabey, *Whistling in the Dark* (London: Sinclair-Stevenson, 1993), p. 11.
[7] B. Roberts, *Nightingale* (London: Reaktion Books, 2021), p. 147.
[8] S. Lee, *The Nightingale* (London: Century, 2020), p. 151.

Then was the lyre of earth beheld,
Then heard by me: it holds me linked;
Across the years to dead-ebb shores
I stand on, my blood thrill restores.
But would I conjure into me
Those issue notes, I must review
What serious breath the woodland drew;
The low throb of expectancy;
How the white mother-muteness pressed
On leaf and meadow-herb; how shook,
Nigh speech of mouth, the sparkle-crest
Seen spinning on the bracken-crook.[9]

---

[9] G. Meredith, *Selected Poems* (London: Elibron Classics, 2005), pp. 45–7.

Bibliography

Adams, H. G. *Favourite Song Birds: Feathered Songsters of Britain.* 1851 (reprinted Delhi: Pranava Books, 2021).
Aristophanes. Translated by S. Halliwell. *Birds and Other Plays.* Oxford: Oxford University Press, 1998.
Aristotle. *De Anima.* London: Penguin, 1986.
Audubon, J. J. *The Audubon Reader.* (ed.) R. Rhodes. New York: Alfred A. Knopf, 2006.
Augiatis, D., and Lander, D. *Radio Rethink: Art, Sound and Transmission.* Banff: Walter Phillips Gallery, 1990.
Baker, J. A. *The Peregrine.* London: William Collins, 2017.
Baldwin, S. *On England, and Other Addresses.* London: Philip Allen, 1926.
Barrell, J. *The Idea of Landscape and the Sense of Place.* Cambridge: Cambridge University Press, 1972.
Bate, J. *John Clare: A Biography.* London: Picador, 2004.
BBC Year Book 1946. London: British Broadcasting Corporation, 1946.
BBC Year Book 1949. London: British Broadcasting Corporation, 1949.
BBC Year Book 1951. London: British Broadcasting Corporation, 1951.
Benson, S., and Montgomery, W. *Writing the Field: Sound, Word, Environment.* Edinburgh: Edinburgh University Press, 2018.
Berger, J. *Understanding a Photograph.* London: Penguin Modern Classics, 2013.
Berry, M., and Brock, D. W. E. *Hunting by Ear.* London: H. F. & G. Witherby, 1937.
Beston, H. *The Outermost House.* London: Pushkin Press, 2019.
Bevis, J. *Aaaaw to Zzzzzd: The Words of Birds.* Cambridge, MA: MIT Press, 2010.
Birkhead, T. *The Wonderful Mr Willughby: The First True Ornithologist.* London: Bloomsbury, 2018.
Bloomfield, R. *Selected Poems of Robert Bloomfield.* (ed.) J. Goodridge and J. Lucas. Nottingham: Trent Editions, 1998.
Bonner, A. *The Songs of the Troubadours.* New York: Schocken Books, 1972.
Bonta, M. M. *Women in the Field: America's Pioneering Women Naturalists.* College Station: Texas A & M University Press, 1992.
Boston, L. M. *The Chimneys of Green Knowe.* Hemingford Grey: Oldknow Books, 1016.
Bull, M., and Back, L. (eds) *The Auditory Culture Reader.* New York: Berg, 2003.
Burroughs, J. *Birds and Poets.* [Reprint] Frankfurt am Main: Outlook, 2019.
Burroughs, J. *The Art of Seeing Things.* (ed.) C. Z. Walker. Syracuse: Syracuse University Press, 2001.

Carey, B., Greenfield, S. and Milne, A. *Birds in Eighteenth-Century Literature: Reason, Emotion, and Ornithology, 1700–1840*. Cham: Palgrave Macmillan, 2020.

Carr, S. *Ode to the Countryside*. London: National Trust Books, 2010.

Carson, R. *Silent Spring*. London: Penguin Books, 2000.

Chaucer, G. (ed.) F. N. Robinson. *The Complete Works of Geoffrey Chaucer*. London: Oxford University Press, 1974.

Clare, J. *The Poems of John Clare*, volumes l and ll. (ed.) J. W. Tibble. London: Dent, 1935.

Clare, J. *Selected Poems and Prose of John Clare*. (ed.) R. Robinson and G. Summerfield. Oxford: Oxford University Press, 1978.

Clare, J. *Bird Poems*. London: Folio Society, 1980.

Clare, J. *The Early Poems of John Clare 1804–1822, Volume ll*. (ed.) E. Robinson and D. Powell. Oxford: Clarendon Press, 1989.

Clare, J. *Selected Letters*. (ed.) M. Storey. Oxford: Oxford University Press, 1990.

Cocker, M., and Mabey, R. *Flora Brittanica*. London: Chatto and Windus, 2005.

Coleridge, S. T. *Coleridge: Poetry and Prose.* (ed.) H. W. Garrod. Oxford: Clarendon Press, 1954.

Coleridge, S. T. *The Works of Samuel Taylor Coleridge*. Ware: Wordsworth Editions, 1994.

Collier, M. *Songs of Place and Time: Birdsong and the Dawn Chorus in Natural History and the Arts*. Manchester: Cornerhouse Publications/Gaia Project Press, 2020.

Cowper, W. *The Poetical Works of William Cowper*. London: Frederick Warne, 1893.

Crabbe, G. *The Poetical Works of George Crabbe*. Edinburgh: Gall and Inglis, 1854.

Darling, L., and Darling, L. *Bird*. London: Methuen, 1963.

Darwin, C. *The Origin of Species*. London: Penguin Classics, 1985.

Darwin, C. *The Voyage of the Beagle*. New York: Meridian Publishing, 1996.

Darwin, C. *The Descent of Man*. London: Penguin Classics, 2004.

Deacon, G. *John Clare and the Folk Tradition*. London: Sinclair Browne, 1983.

Dee, T. *Greenery: Journeys in Springtime*. London: Jonthan Cape, 2020.

Dickinson, E. *The Complete Poems of Emily Dickinson*. (ed.) T. H. Johnson. London: Faber and Faber, 1970.

Doer, A. *All the Light We Cannot See*. London: Fourth Estate, 2015.

Donne, J. *The Complete Poems of John Donne*. London: Dent, 1985.

Drabble, M. *A Writer's Britain*. London: Thames and Hudson, 1979.

Drayton, M. *Polyolbion*. London: John Russell Smith, 1876.

Edwards, P. *Call Me at Dawn*. Ipswich: East Anglian Daily Times, 1948.

Emerson, R. W. *Nature and Selected Essays*. London: Penguin Classics, 2003.

Empiricus, S. Translated by R. G. Bury. *Outlines of Pyrrhonism*. Lanham, MA. Prometheus Books, 1990.

Ewart-Evans, G. *Spoken History*. London: Faber and Faber, 1987.

Foster, C. *Being a Beast*. London: Profile Books, 2016.

Gillin, E. J. *Sound Authorities: Scientific and Musical Knowledge in Nineteenth-Century Britain*. Chicago: University of Chicago Press, 2022.

Greenblatt, S. et al. (ed.) *The Norton Anthology of English Literature*, 8th edition, vol. 2. New York: Norton, 2006.

Grey, E (Viscount Grey of Falloden), *The Charm of Birds*. London: Hodder and Stoughton, 1929.

Grimshaw-Aagaard, M., Walther-Hansen, M. and Knakkergaard, M. *The Oxford Handbook of Sound and Imagination, Volume 1*. Oxford: Oxford University Press, 2019.

Guida, M. *Listening to British Nature: Wartime, Radio, and Modern Life, 1914–1945*. New York: Oxford University Press, 2022.

Gwilym, D. A. *Dafydd ap Gwilym: Fifty Poems*. London: The Honourable Society of Cymmrodorion, 1942.

Gwilym, D. A. *Dafydd ap Gwilym: A Selection of Poems*. London: Penguin Books, 1985.

Hardy, T. *Collected Poems*. London: Macmillan, 1979

Hardy, T. *Under the Greenwood Tree*. London: Penguin Books, 2004.

Harrison, B. *The Cello and the Nightingales*. London: John Murray, 1985.

Haskell, D. G. *The Songs of Trees: Stories from Nature's Great Connectors*. New York: Viking, 2017.

Haskell, D. G. *Sounds Wild and Unbroken*. London: Faber & Faber, 2022.

Holden, P., and Cleeves, T. *The RSPB Handbook of British Birds*. London: Bloomsbury, 2014.

Hooker, J. 'Richard Jefferies: The Art of Seeing'. In *Writers in a Landscape*, edited by J. Hooker, Chapter 2, pp. 16–38. Cardiff: Cardiff University Press, 1996.

Hooker, J. *Word and Stone*. Bristol: Shearsman Books, 2019.

Hooker, J. 'The Tree of Life: Explorations of an Image'. In *Art of Seeing: Essays on Poetry, Landscape Painting and Photography*, edited by J. Hooker, Chapter 6, pp. 95–110. Swindon: Shearsman Books, 2020.

Hopkins, G. M. *The Journals and Papers of Gerard Manley Hopkins*. (ed.) H. House and G. Storey. London: Oxford University Press, 1959.

Hopkins, G. M. *The Poetical Works of Gerard Manley Hopkins*. (ed.) N. H. Mackenzie. Oxford: Clarendon Press, 1990.

Hudson, W. H. *A Traveller in Little Things*. London: Dent, 1921.

Hudson, W. H. *Nature in Downland*. London: Dent, 1923.

Hudson, W. H. *Birds and Man*. London: Duckworth, 1927.

Huxley, J., and Koch, L. *Animal Language*. London: Country Life, 1938.

Jamie, K. *Surfacing*. London: Sort of Books, 2019.

Jefferies, R. *Nature Near London*. London: Chatto & Windus, 1908.

Jefferies, R. *The Story of My Heart*. London: Longman Green, 1936.

Jefferies, R. *The Life of the Fields*. London: Lutterworth Press, 1947.

Jefferies, R. *Field and Hedgerow: The Last Essays of Richard Jefferies*. London: Lutterworth Press, 1948.

Jefferies, R. *The Open Air*. London: Lutterworth Press, 1948.

Jefferies, R. *Landscape with Figures: An Anthology of Richard Jefferies' Prose*. (ed.) R. Mabey. London: Penguin Books, 1983.

Keats, J. *Poetical Works*. London: Oxford University Press, 1967.

Kehew, R. (ed.). *The Lark in the Morning: The Verses of the Troubadours*. Chicago: University of Chicago Press, 2005.

Koch, L. *Memoirs of a Birdman*. London: Phoenix House, 1955.

Koch, L. *The Encyclopaedia of British Birds*. London: Waverley Book, 1955.

Koch, L. *Bird Song*. (Talking Book). London: Talking Book/Methuen, 1960.

Krause, B. *The Great Animal Orchestra*. London: Profile Books, 2012.

Krause, B. *Wild Soundscapes: Discovering the Voice of the Natural World*. New Haven: Yale University Press, 2016.

Leach, E. E. *Sung Birds: Music, Nature and Poetry in the later Middle Ages*. Ithaca and London: Cornell University Press, pp. 40–1.

Lee, S. *The Nightingale: Notes on a Songbird*. London: Century, 2020.

Leighton, A. *Hearing Things: the Work of Sound in Literature*. Cambridge, MA: Belknap Press of Harvard University Press, 2018.

Levi, P. *The Noise Made by Poems*. London: Anvil Press, 1977.

Longley, M. *Sidelines: Selected Prose 1962-2015*. London: Enitharmon Press, 2017.

Lonsdale, R. (ed.). *Eighteenth Century Women Poets*. Oxford: Oxford University Press, 1989.

Lonsdale, R. *The New Oxford Book of Eighteenth Century Verse*. Oxford: Oxford University Press, 2009.

Mabey, R. *Whistling in the Dark: In Pursuit of the Nightingale*. London: Sinclair-Stevenson, 1993.

Mabey, R. *The Barley Bird: notes of the Suffolk Nightingale*. Woodbridge: Full Circle Editions, 2010.

Mabey, R. *Turning the Boat for Home: A Life Writing About Nature*. London: Chatto and Windus, 2019.

Maxwell, G. *On Poetry*. London: Oberon Books, 2012.

Maybury, T. *Coleridge and Wordsworth in the West Country*. Stroud: Alan Sutton Publishing, 1992.

McCarthy, M., Mynott, J. and Marren, P. *The Consolation of Nature: Spring in the Time of Coronavirus*. London: Hodder Studio, 2020.

McNulty, Dara. *Diary of a Young Naturalist*. Toller Fratrum: Little Toller Books, 2020.

Menke, R. *Literature, Print Culture, and Media Technologies, 1880-1900*. Cambridge: Cambridge University Press, 2019.

Meredith, George. *Selected Poems*. Boston, MA: Elibron Classics, 2005.

Milton, J. *John Milton: The Complete Poems.* (ed.) J. Leonard. London: Penguin Books, 1998.
Morton, E. S., and Page, J. *Animal Talk: Science and the Voices of Nature.* New York: Random House, 1992.
Motion, A. *Keats.* London: Faber and Faber, 1997.
Muir, J. *Journeys in the Wilderness: A John Muir Reader.* Edinburgh: Birlinn, 2017.
Mynott, J. *Birdscapes: Birds in our Imagination and Experience.* Princeton: Princeton University Press, 2009.
Nicholson, E. M., and Koch, L. *Songs of Wild Birds.* London: H.F & G. Witherby, 1937.
Nicholson, E. M., and Koch, L. *More Songs of Wild Birds.* London: H.F & G. Witherby, 1937.
Oliveros, P. *Sounding the Margins: Collected Writings 1992–2009.* Kingston, NY: Deep Listening, 2010.
Ovid. Translated by A. D. Melville. *Metamorphoses.* Oxford: Oxford University Press, 1986.
Piette, A. *Remembering and the Sound of Words.* Oxford: Clarendon Press, 2004.
Pliny the Elder. Translated by J. J. Healy. *Natural History: A Selection.* London: Penguin Books, 2004.
Pope, A. *Collected Poems.* London: Dent, 1969.
Quiller-Couch, A. (ed.) *The Oxford Book of English Verse.* Oxford: Clarendon Press, 1931.
Raven, C. E. *John Ray: Naturalist.* Cambridge: Cambridge University Press, 1950.
Renehan, E. J. *John Burroughs: An American Naturalist.* Post Mills, VE: Chelsea Green, 1992.
Richards, J. *Voices and Books in the English Renaissance.* Oxford: Oxford University Press, 2019.
Roberts, B. *Nightingale.* London: Reaktion Books, 2021.
Robinson, E., and Fitter, R. *John Clare's Birds.* Oxford: Oxford University Press, 1982.
Rosenberg, I. *Collected Poems.* London: Chatto and Windus, 1974.
Rothenberg, D. *Why Birds Sing.* London: Penguin, 2006.
Schafer, R. M. *The Soundscape: Our Sonic Environment and the Tuning of the World.* Rochester: Destiny Books, 1994.
Sellers, P. *The Best of Sellers.* London: EMI Records MRS 5157, 1958.
Shelley, P. B. *Selections from Shelley's Poems.* (ed.) D. Welland. London: Hutchinson Educational, 1961.
Shepherd, N. *The Living Mountain.* Edinburgh: Canongate Publishing, 2014.
Simms, E. *Wildlife Sounds and Their Recording.* London: Paul Elek, 1979.
Smith, M. M. *Hearing History: A Reader.* Athens/London: University of Georgia Press, 2004.
Smyth, R. *An Indifference of Birds.* Axminster: Uniform Books, 2020.
Stein, G. *Look at Me Now and Here I Am.* London: Peter Owen, 2004.

Stewart, S. *Reading Voices: Literature and the Phonotext*. Berkeley: University of California Press, 1990.

Street, S. *The Poetry of Radio: The Colour of Sound*. Abingdon, Routledge, 2012.

Street, S. *Sound Poetics: Interaction and Personal Identity*. Cham: Palgrave Macmillan, 2017.

Street, S. *Sound at the Edge of Perception*. Singapore: Palgrave Macmillan, 2019.

Swann, K. H. *A Dictionary of English and Folk-Names of British Birds*. London: Read Books, 2013.

Thomas, E. *Richard Jefferies: His Life and Work*. Boston: Little, Brown, 1909.

Thomas, E. *Letters from Edward Thomas to Gordon Bottomley*. (ed.) R. G. Thomas. London: Oxford University Press, 1968.

Thomas, E. *The South Country*. London: Hutchinson, 1994.

Thomas, E. *Edward Thomas: The Annotated Collected Poems*. (ed.) E. Longley. Tarset: Bloodaxe Books, 2008.

Thomson, J. *The Poetical Works of James Thomson*. Edinburgh: William P. Nimmo, 1883.

Thoreau, H. D. *The Journals, 1837–1861*. (ed.) D. Searls. New York: New York Review of Books, 2009.

Thoreau, H. D. *Walden*. London: Penguin Books, 2016.

Thoreau, H. D. *Walking*. Los Angeles: Enhanced Media, 2017.

Tomalin, R. *W. H. Hudson: A Biography*. London: Faber and Faber, 1981.

Wallace, A. R. *My Life: A Record of Events and Opinion*. London: Pantianos Classics, 1908.

Wallace, A. R. *The Malay Archipelago*. London: Penguin Classics, 2014.

Walton, I. *The Compleat Angler*. Oxford: Oxford University Press, 1974.

White, G. *The Natural History and Antiquities of Selborne*. London: Walter Scott, 1887.

White, G. *The Natural History and Antiquities of Selborne*. London: Dulcimus Books, 1974.

White, G. *Gilbert White's Year*. (ed.) J. Commander. Oxford: Oxford University Press, 1982.

White, G. *The Journals of Gilbert White, Volume Three: 1784–1793*. (ed.) F. Greenoak. London: Century, 1985.

Williams, R. *The Country and the City*. London: Vintage, 2016.

Willughby, F., and Ray, J. *Ornithology*. London: John Martyn, 1676.

Winn, R. *The Salt Path*. London: Penguin Books, 2018.

Wood, J. G. *Out of Doors: A Selection of Original Articles on Practical Natural History*. London: Longmans, Green, 1874.

Wordsworth, D. *The Grasmere and Alfoxden Journals*. Oxford: Oxford University Press, 2008.

Wordsworth, W. *The Poetical Works of William Wordsworth*. London: Henry Frowde/Oxford University Press, 1909.

Wordsworth, W. *Prefaces to the 'Lyrical Ballads.'* London: Thomas Nelson and Sons, 1937.

Wordsworth, W. *The Prelude, or Growth of a Poet's Mind.* Oxford: Clarendon Press, 1959.

Index

Albin, Eleazar 97–8
   A Natural History of English Songbirds 97–8
Aldburgh, (Village in Suffolk) 101–3
Amherst, Massachusetts 147
Anderson, Robert 59
   'To a Redbreast' 59–60
Aristophanes 82
   The Birds 82–3
Aristotle 55
Attenborough, David 195
Audubon, John James 158–9
   The Birds of America 158–9

Bach, J. S. 192
   Prelude and Fugue in A Minor 192
Baker, John Alec 190–2
   The Peregrine 190–2
Baldwin, Stanley 63
Barrington, Daines 105
Beethoven, Ludwig van 8
   'Pastoral' Symphony 8
Bell, Alexander Graham 184
Berger, John 200
Berliner, Emile 184
Berry, Francis 83
Berry, Michael 41, 42
Beston, Henry 195
   The Outermost House 195
Bevis, John 196
   Aaaaw to Zzzzd: The Words of Birds 196
Bewick, Thomas 131
birds
   bittern 46
   Blackbird 15, 33, 50 (Merle and Woosell in Drayton), 86
   blackcap 36
   blue Jay 69
   bobolink 150, 159
   Brown Thrasher 5, 69
   bullfinch 23

chaffinch 23, 114
chanticleer (farmyard cock) 114
chickadee 157
chiff-chaff 22, 25
chough 88
cock (*see also* Chanticleer) 119
cormorant 131
crake
crane 188
crow 87, 89, 139, 145
cuckoo 7–8, 9, 74, 170, 171
curlew 36, 43, 64, 195
dove (Turtle) 87, 117
duck 74, 139
eagle 88
gannet 14
goldcrest 188
goldfinch 23, 36, 49–50
goose 74, 87, 88
Great Bird of Paradise 167
greenfinch 23
groundlark 145
heron 145, 196
humming bird 149
Indian Sharma Bird 5
kingfisher 172
lark (*see also* skylark) 28–30, 32, 59, 60, 64, 87, 96, 113
linnet 117
lyrebird 5
magpie 33, 69
mistle-thrush 36, 50
nightingale 2–3, 5, 15, 25, 27, 29, 30, 31, 55–6, 72–3, 74, 78, 79, 80, 81, 82, 86, 88, 96–7, 100, 115, 136–7, 138–9, 201–5
nightjar (goatsucker) 105–6, 134–5, 159
ostrich 167
owl 64, 89, 104 (Gilbert White on musical pitch) 107, 131, 181
   barred 158

screech 163
tawny, 44
white 107
parrot 68, 69
peregrine 190–2
pewit 139
pigeon 139
pipit 14 (referred to as titlark) 181
purple finch 157
raven 50, 88, 89, 117
redshank 93, 195
robin (Robin Redbreast) 23, 58–60, 70, 127, 145, 193
rook 50, 114
sedge Warbler 186
snipe 14
skylark (*see also* Lark) 14, 30, 36, 127, 170, 180, 188, 191–2
song thrush 5, 50, 60, 69, 80, 81, 114, 117, 157–8 (Hermit Thrush) 173
sparrow 50, 62, 117
starling 50, 68
swallow 50, 127, 173
swan 82–3
wagtail 114
willow Warbler 23
woodlark 29 (n), 36, 108, 114, 145, 170
wren 58, 136
Bloomfield, Robert 99–100, 131–2, 141
  'The Farmer's Boy' 99–100, 132
Boston, Lucy M. 194
  The Chimneys of Green Knowe 194
Bridgeman, Charles 113
British Broadcasting Company/ British Broadcasting Corporation (BBC) 2, 31, 45, 46, 50, 181, 195
  European Service 45
  Home Service 45
B.I.R.S. (British Institute for Recorded Sound) 48
BBC Year Book 53
British Library 48
British Library – Sounds 48
Brock, D. W. E. 41, 42
Brown, Lancelot 'Capability' (n) 113
Buchan John 95
Bunting, Basil 69
Bunyan, John 96

The Pilgrim's Progress 96–7
Burghley House 130
Burroughs, John 155–9, 164
  Birds and Poets 155
  The Art of Seeing Things 159

Carroll, Lewis (Charles Lutwidge Dodgson) 148
  Alice in Wonderland 148
Carson, Rachel 198
Chamberlain, Neville 40, 48
Chaucer, Geoffrey 15, 68, 72–6, 78, 81, 82, 202
  The Canterbury Tales (Prologue) 73–4
  The Legend of Good Women 202
  The Parlement of Foules 74–6, 82
  Troilus and Criseyde 72–3
Chaytor, H. J. 77
Chelmsford, Essex 190–1
Chester, Charlie 195
Clare, John 13, 64, 103, 113, 116, 129–47, 153, 184
  'A Sunday with Shepherds and Herdboys' 140
  'Dyke Side' 135–6
  'Emmonsail's Heath in Winter' 145
  'Emmonsale's Heath' 144–5
  'Honest John' 146
  'Narrative Verses Written after an Excursion to Burghley Park' 130
  'Pleasant Sounds' 145
  Poems Descriptive of Rural Life and Scenery 132
  'Spring' 129
  'Sunday' 140
  'The Flight of Birds' 139
  'The Nightingale' 138–9
  'The Nightingale's Nest' 136–7
  The Village Minstrel 130
  'The Wren' 136
Cocker, Mark 68
Coleridge, Samuel Taylor 109, 113, 115–23, 125, 127, 129, 169, 202
  'Answer to a Child's Question' 116–17
  'Christabel' 119
  'Lines Written in the Album at Elbingerode, in the Hartz Forest' 121

'The Nightingale' 115–16
'The Rhyme of the Ancient Mariner' 123
Conan Doyle, Sir Arthur 66
  The Adventure of the Lion's Mane 66
Concord, Massachusetts 160–4
Constable, John 134
Country Life Magazine 43
Cowley, Abraham 141
Cowper, William 105, 109
  The Task 105, 109
Crabbe, George 101–3
  The Borough 102–3
  The Village 101–2

Dafydd ap Gwilym 79–81, 97, 172
  'The Woodland Mass' 81
D'Alvernhe, Peire 78–9
  'Nightingale, for me take flight' 79
Daniel, George 70
Darby, Abraham 112
Darwin, Charles 21, 75, 165, 167–8, 174, 176, 179, 180, 182, 197
  The Descent of Man 75, 168, 180
  The Origin of Species 165, 182
  The Voyage of the Beagle 167
Davidson, John 64–5
  'Spring Song' 64–5
Debussy, Claude 16
  Syrinx (Musical work) 16
Dee, Tim 188
  Greenery: Journeys in Springtime 188
De la Primaudaye, Pierre 84
Delius, Frederick 9
  On Hearing the First Cuckoo in Spring 9
De Martinville, Edward-Leon Scott 183–4
Descartes, Rene 92
  Discourse on Method 92
Dickinson, Emily 147–51, 153, 154, 159, 164, 174, 176
  'Great Streets of Silence Led Away' 148
  'I Heard a Fly buzz – when I died' 151
  'I never Saw a Moor' 147
  'I Think that the Root of the Wind is Water' 149
  'Some Keep the Sabbath Going to Church' 150
  'The Saddest Noise, the Sweetest Noise' 150–1

'There is no Silence in the Earth – so silent' 151
'Within my Garden Rides a Bird' 149
Donne, John 76, 95
  'An Epithalanium' 76
Drayton, Michael 84–6
  Poly-Olbion 84–6
Dvorak, Antonin 2
  Songs my Mother Taught Me 2
Duck, Stephen 61–2
  'The Thresher's Labour' 61–2
Duns Scotus 172

Edison, Thomas 4, 155, 184
Edwards, Percy 195–6
  Call Me at Dawn 195
Elgar, Edward 2
  Cello Concerto 2
Eliot, T. S. 65, 202
  The Waste Land 202
Elliott, Ebenezer 111
  'Steam at Sheffield' 111
Emerson, Ralph Waldo 152, 160, 161, 169
Epstein, Jacob 177
Ewart Evans, George 55

Farina, Carlo 10
  Capriccio stravagante 10
Finch, Anne, Countess of Winchelsea 100–1, 113
  'To the Nightingale' 100
  'A Pastoral Dialogue between Two Shepherdesses' 101, 113
Firestone, Harvey 155
First Sounds Project 184
Fisher, James 49, 186, 195
Ford, Henry 155
Foster, Charles 33
  Being a Beast 33
Freud, Sigmund 167
Frost, Robert 25, 122, 185
  'Stopping by Woods on a Snowy Evening' 25

Gay, John 141
George V (King) 3
Gillard, Frank 45, 53
Gladstone, W. E. 4

Goldsmith, Oliver 57–8
  'The Deserted Village' 57–8
Gould, Maude 3–4, 195
Gray, Thomas 104–5
  'Ode on the Spring' 104
  'Elegy Written in a Country Churchyard' 104
Grey, Edward 188–9
  The Charm of Birds 188
Guida, Michael 29, 30

Hardy, Thomas 1–2, 4, 27, 53, 56, 58, 178, 201, 203
  'The Spring Call' 56–7
  Under the Greenwood Tree 1, 27, 201
Harrison, Beatrice 1–6, 27, 31, 181, 197
Harvey, Jonathan 9
  Bird Concerto with Piano Song 9
Haskell, David 34, 187
Hawkins, Desmond 43, 49, 50, 53
Haydn, Joseph 10
  Symphony no. 83 (The Hen) 10
Hazlitt, William 99
Helmholtz, Hermann von 168
Helpston, (Village in Northamptonshire) 129, 130, 133
Hesiod 112, 113
H.F. & G. Witherby (publishers) 39–42
Higginson, Thomas Wentworth 147–8, 174
Higher Bockhampton (Dorset) 1, 27
Holden, Edith 63
  The Country Diary of an Edwardian Lady 63
Holy Island (Cumbria) 8
Hooker, Jeremy 116, 134, 175, 199
  'Somewhere, a Blackbird' 199
Hopkins, Gerard Manley ix, 15, 169–73, 179
  'Spring' 172–3
  'The Kingfisher' 172
  'The Sea and the Skylark' ('Walking by the Sea') 170–1
  'The Woodlark' 170
Horsey Mere (Suffolk) 46
Hosking, Eric 49
Hudson, William Henry (W.H.) 25, 176–81
  A Traveller in Little Things 178–9
  Birds and Man 178, 180

Hunt, Leigh 114
Huxley, Julian 25, 36, 39, 45, 47, 49

Jamie, Kathleen 26
  Surfacings 26
Jefferies, Richard 13, 16, 23, 24, 25, 173–7, 179, 180, 184, 185
  After London 23–4, 173
  'Hours of Spring', in *Field and Hedgerow* 181
  Nature Near London 173
  The Life of the Fields 174–5
  The Open Air 174
  The Story of My Heart 173, 180

Kearton, Cherry 5
Keats, John 15, 31, 123–7, 129, 134, 138, 185, 197, 202, 203
  'Ode on a Grecian Urn' 125–6, 197
  'Ode to Autumn' 126–7
  'Ode to a Nightingale' 124–5, 203
  'The Eve of St Agnes' 202
Koch, Ludwig 5, 15, 24–5, 35–51, 53, 65, 187, 199
  Animal Language 36, 43
  Hunting by Ear 41–2
  More Songs of Wild Birds 24, 43
  Oiseaux Chanteurs de Laeken 44
  Songs of Wild Birds 24–5, 40
  Sound-Books 35
  The Language of Birds 47
Krause, Bernie 192–3, 199

Lane, William 59
Leighton, Angela 31
Levi, Peter 135
Liddell, Alice 148
Lindisfarne Gospels 8
Longley, Michael 136–7
Lyrical Ballads 115, 122–3

Mabey, Richard 3, 15, 29 (n), 107, 108, 113, 175, 203
  The Barley Bird 15
  Whistling in the Dark 203
MacDiarmid, Hugh 65
'Madame Saberon' *see* Gould, Maude

Index

Mahler, Gustav 8
   Symphony No. 1 ('Titan') 8–9
McNulty, Dara 14, 198
Maxwell, Glyn 198
Meredith, George 11–12, 204
   'Night Frost in May' 204–5
   'The Lark Ascending' 11–12
Messiaen, Olivier 9, 29 (n)
   Catalogue d'oiseaux 9
   Le merle noir 9
   Reveil des oiseaux 9
   Turangalila-Symphonie 9
Miller, Max 195
Milton, John 141–2
   'L'Allegro' 141
   'Il Penseroso' 141–2
Mitchell, Denis 55
Morecambe and Wise 195
Mozart, Leopold 10
   Toy Symphony 10
Mozart, Wolfgang Amadeus 67
Muir, John 151–5, 164
   The Story of my Boyhood and Youth 154–5
   The Yosemite 152–3, 153–4
Murray Shafer, R. 72
Mynott, Jeremy 3, 22

National Sound Archive, The 48
Newton, Alfred 93
   Dictionary of Birds 93
Newton, Isaac 92, 103, 129
   Principia Mathematica 92
Nicholson, E. M. 24, 35, 40
Northampton Asylum (St Andrew's Hospital) 133, 145, 146

Oliveros, Pauline 27, 120, 188, 189, 193
   Sounding the Margins 188
Olney (Town in Buckinghamshire) 109
Ovid 202

Parker, Charles 55
Pennant, Thomas 105
'Philomena' legend 202–3
Pliny the Elder 67, 71
   Natural History 71
Plowright, Piers 28

Pope, Alexander 61, 91–2, 101
   An Essay on Criticism 91–2
   'Sound and Sense' 91–2
Pope Pius IX 169
   The Syllabus of Errors 169
Poulson, Valdemar 6 (n)
Pound, Ezra 77
Prokofiev, Serge 10
   Peter and the Wolf 10
Pyrrho of Elis 71–2

Rautavara, Einojuhani 10
   Cantus Arcticus: Concerto for Birds and Orchestra 10–11
Ravel, Maurice 10
   Sites Auriculaires 10
Ray, John 93–4, 103
   A Collection of English Words 94
   Ornithology (with Francis Willughby) 93–4
Reich, Karl 5
Reith, John 2
Repton, Humphry 113
Respighi, Ottorino 9
   The Birds 9
Rimsky-Korsakov, Nikolai 2
   Chant Hindou 2
Robinson, Marilynne 26
   Housekeeping 26
Robinson, Mary 99
   'London's Summer Morning' 99
Rosenberg, Isaac 28–30
   'Returning, We Hear the Larks' 28–30
Rothenberg, David 197
Roxbury, Delaware County, New York 155, 157
Royal Society, The 93
R.S.P.B. (Royal Society for the Protection of Birds) 176
Rudel, Jaufre 78
   'The Nightingale' 78
Rushton, Edward 60
   'To a Redbreast …' 60

Saint-Saëns, Camille 10
   The Carnival of the Animals 10
Sample, Geoff 193, 199

Saul, Patrick 48
Scott, Peter 49
Scott, Walter 77
  'The Lay of the Last Minstrel' 77
  'The Troubadour' 77
Selborne (Village in Hampshire) 105–9, 157
Sellers, Peter 44–5, 195
Seward, Anna 111–12
  'Sonnet: To Colebrooke Dale' 112
Sextus Empiricus 71
  Outlines of Pyrrhonism 71–2
Shakespeare, William 15, 83–90, 202–3
  A Midsummer Night's Dream 88
  As You Like It 86, 87, 88, 203
  A Winter's Tale 87
  Henry VI Part III 89
  King Lear 91
  The Merchant of Venice 87
  The Rape of Lucrece 202–3
  The Tempest 89
  The Two Gentlemen of Verona 88
  The Winter's Tale 87
  Titus Andronicus 88–9, 202
Shelley, Percy Bysshe 127
  'To a Skylark' 127
Shepherd, Nan 189
  The Living Mountain 189
Sibelius, Jean 10
  The Swan of Tuonela 10
Simms, Eric 5, 47, 49
Soper, Tony 195
Southey, Robert 131
Stanwood, Cordelia 156–7, 164
Stein, Gertrude 73
Stevens, Wallace 65
Strauss, Max 38
Swift, Jonathan 61

Takemitsu, Tōru 10
  A Flock Descends into the Pentagonal Gardens 10
Taylor & Hessey (publishers) 130
Taylor, John 130
Thatcher, Margaret 63
Theocritus 112, 113
Thomas, Edward 16, 122, 182–3, 184–7
  'Old Man' 186

  Richard Jefferies: His Life and Work 182–3
  'Sedge Warblers' 186
  The South Country 184–5
  'The Unknown Bird' 185
Thomson, James 103, 129, 130, 140
  The Seasons 103, 129, 130–1
  'To the Memory of Isaac Newton' 103
Thoreau, Henry David 13, 16, 160–5, 169, 181
  Journal 160–1, 163, 165
  Kalendar 160
  Walden 162, 163
  Walking 164
Tibble, J. W. 137–8
Troubadours 77–81

Vaughan William, Ralph 11, 12
  The Lark Ascending 11
Ventadorn, Bernart de 77–8
Vesey-Fitzgerald, Brian 49
Virgil 112
Vivaldi, Antonio 10
  The Four Seasons 10–11

Wagner, Cosima 38
Wagner, Siegfried 38
Wallace, Alfred Russel 167–8, 174, 179
  The Malay Peninsula 167–8
  My Life: A Record of Events and Opinion 179
Walton, Izaak 15, 95–7, 98
  The Compleat Angler 95–7
Watson, Chris 8, 187–8, 199
Weldon Champneys, W. (Dean of Lichfield) 169
White, Gilbert 13, 15, 25, 95, 105–9, 113, 157, 181
  Journal 107, 108–9
  The Natural History and Antiquities of Selborne 105–8
Whitman, Walt 157, 169
  'When Lilacs Last in the Doorway Bloom'd' 147
Wilkins, John 93
William de Wycombe 7
Willughby, Francis 93–4, 103

Ornithology (With John Ray) 93–4
Winn, Raynor 190
  The Salt Path 190
Witherby, George 41
Witherby, Harry 39–40
Wood, J. G. 66
  Out of Doors 66
Wordsworth, Dorothy 115–19, 141
  Alfoxton Journal 115
  Grasmere Journal 117–19, 141

Wordsworth, William 13, 15, 30, 109, 113, 115–24, 125, 126, 127, 130, 131, 134, 137, 142, 169
  'Lines Written Above Tintern Abbey' 109, 123, 124
  'On the Power of Sound' 119–20, 124, 126
  'The Idiot Boy' 117, 126
  The Prelude 30–1, 109, 118
  'The Thorn' 122

www.ingramcontent.com/pod-product-compliance
Lightning Source LLC
Chambersburg PA
CBHW052038300426
44117CB00012B/1872